ISBN 978-1-332-09441-7
PIBN 10283815

1 MONTH OF FREE READING

at

www.ForgottenBooks.com

By purchasing this book you are eligible for one month membership to ForgottenBooks.com, giving you unlimited access to our entire collection of over 1,000,000 titles via our web site and mobile apps.

To claim your free month visit:

www.forgottenbooks.com/free283815

Prepared in connection
with Project NA-195
under Contract OEMsr-945

Approved for Distribution
Warren Weaver
Chief, Applied Mathematics Panel

Distribution List

AMP Report 137.1R

Copy No.

1 - 7	Office of Executive Secretary, OSRD
8 - 34	Liaison Office, OSRD
35	Office of Research and Inventions Att: Lt. J. H. Wakelin
36	Commanding General, AAF Att: T. von Kármán
37 - 40	Chief, Bureau of Aeronautics 1 Comdr. J. S. Warfel 1 Lt. F. A. Parker 1 E. S. Roberts
41 - 45	Bureau of Aeronautics Representative in Los Angeles Att: Aerojet Corp.
46 - 47	U. S. Naval Engineering Experimental Station, Annapolis 1 Lt. Patton 1 Ensign L. B. Edelman
48 - 54	Chief, Bureau of Ordnance 1 R. S. Burington 1 R. J. Seeger 1 Comdr. Levering Smith, Re2d 1 Comdr. W. A. Walter, Re3 1 Capt. D. P. Tucker Re4g 1 Capt. E. M. Crouch Re6
55 - 59	Commanding General, AAF Materiel Command, Wright Field 1 Lt. Col. P. F. Nay 1 Major J. P. AuWerter 1 Capt. J. Healy, Armament Laboratory 1 Lt. R. C. Bogert 1 Bell Aircraft Corp.
60 - 61	J. E. Jackson, Chairman's Office, NDRC 1 Bumblebee Project
62	Guided Missiles Committee, JCS

63 - 66   Colonel L. A. Skinner, Ordnance Department Representative, C.I.T.
    1 C. Millikan
    1 L. G. Dunn
    1 J. V. Charyk

67 - 69   Office of the Chief of Ordnance
    1 Colonel G. W. Trichel
    1 Major J. F. Miller, Jr.
    1 H. M. Morse

70   Commanding General, AAF Proving Ground Command, Eglin Field, Fla.
    Att: Brig. Gen. Grandison Gardner

71 - 72   Ordnance Research Center, Aberdeen Proving Ground
    1 Technical Library
    1 R. J. Walker

73 - 74   F. L. Hovde, Chief, Division 3
    1 E. C. Watson
    1 J. B. Rosser

75   J. T. Tate, Chief, Division 6

76   W. R. Kirner, Chief, Division 9
    Att: J. W. Williams

77   W. Weaver, Chief, Applied Mathematics Panel

78   T. C. Fry, Acting Chief, Applied Mathematics Panel

79 - 80   R. Courant

81   O. Veblen

82   G. C. Evans

83   S. S. Wilks

84   M. Rees

PREFACE

The present report, by the New York University Group of the Applied Mathematics Panel, was undertaken in connection with Project No. NA-195 (Bureau of Aeronautics, Navy Department). It was carried out, under the general guidance and editorship of Dr.J.J.Stoker, by M. Shiffman, D. C. Spencer, B. Friedman, E. Bromberg and E. Isaacson in consultation with Dr. F. Zwicky of the Aerojet Corporation.

The report contains a mathematical analysis of a variety of problems concerning the mechanics of the exhaust stroke in a hydropulse motor. A broad and basic approach to the problems was motivated to a considerable extent by the desire to avoid too specific assumptions with regard to the engineering aspects of the hydropulse development.

R. Courant
Director of Research
Contract OEMsr-945

TABLE OF CONTENTS

All graphs associated with a section are to be
found immediately following that section.

<u>Ideal Mechanical Performance Characteristics</u>

<u>of the Hydropulse</u>

<u>Section 1. Introduction and summary.</u>

In this report we are interested in certain aspects
of the problems which arise in dealing with jet propulsion
devices to be used in water -- in other words, with the
counterparts in water of the jet devices in air, such as
rockets, the athodyd, the buzz bomb motor, etc. The
manner in which such devices function can be described in
a general way as follows: The propulsive element is a duct

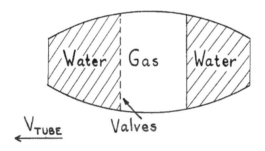

Figure 1.1

open at both ends (as indicated in Figure 1.1). The duct
is assumed to be traveling through the water with a certain
velocity V (presumably it is propelling a ship, a torpedo,
or other craft). At a certain instant a chemical is
injected into the duct which reacts with the water to produce
a volume of gas, as indicated in the figure. The gas, on
expanding, imparts momentum to the water and a thrust in the
forward direction will result if, for example, the part of

CONFIDENTIAL

the duct in front of the gas is sealed off by closing valves.
If the forward velocity is high enough and if the intake
part of the duct is a properly designed expanding nozzle,
valves may not be needed to insure a forward thrust and the
fuel could be injected continuously. Such a device would
correspond to the athodyd in air. However, at moderate
velocities V, it is likely that valves would be necessary
for efficient operation of such a duct device. Of course,
sustained operation of the device with valves requires
the valves to open at certain times so that the expansion
chamber can be refilled with water after the gas has expanded
and pushed a predetermined amount of water out of the tube.
Once the "expansion chamber" has been refilled, the valves
would be closed, fuel would be injected, and the cycle
would repeat.

In what follows we assume that valves will be provided,
but we do not assume that we know when and how they function --
whether they open automatically when the gas pressure reaches
a certain value (the hydro-resonator), or are controlled by
some independent mechanism (hydropulse).[*] Our investigations
have therefore been confined for the most part to those
phenomena which can be treated without knowing in detail

---

[*] This terminology we have taken from the report of F.
Zwicky: Remarks on the Basic Theory of the Hydropulse as
a Propulsive Power Generator, Aerojet Report No. R-41
(Conf.). In fact, most of the work to be discussed in
what follows was undertaken at the suggestion of F. Zwicky
and in consultation with him. Two other reports of interest
for the mechanical performance of the hydropulse have
appeared: 1) The Solution of the Differential Equation
for the Hydropulse (Conf.), Aerojet Report No. R-43, by
K. Foelsch and Knox T. Millsaps, 2) A Preliminary Study
of the Hydropulse (Conf.), Jet Propulsion Note No. 19,
Bureau of Aeronautics Project, U.S. Naval Eng. Exp.
Station, Annapolis, Md., written by Ensign L. B. Edelman.

just how the complete cycle of operations will be carried out.
Our calculations should be helpful in making decisions on
this point. Roughly speaking, what we have done is the
following: We have ignored the part of the cycle concerned
with the refilling of the "expansion chamber", (except in
Section 12) and concentrated our attention on that part of
the cycle during which the gas is expanding and pushing water
out of the tube. We consider the motion up to the time when
the gas has expanded down to the exit pressure, and define
as the time of stroke -- or simply the stroke -- the time
required for such an expansion. In carrying out the calculations certain simplifying assumptions are made which are
discussed in Section 2, immediately following this introduction. Briefly speaking, these assumptions lead to a
"hydraulic theory", in which the water is treated as
incompressible and the velocity distribution over any cross-
section of the tube is replaced by the average over the
section. In addition, the gas is assumed to be uniform in
state throughout its volume at each instant of time, and to
expand adiabatically in such a way that the interface between
gas and water is at all times a plane surface at right angles
to the axis of the tube. The pressure at the exit of the
tube is assumed to be constant throughout the stroke. On the
basis of these assumptions we study the motion of the water
in the tube, the impulse delivered over the stroke, and the
time of stroke -- that is, we study the mechanics of the
hydropulse operation, but leave aside the chemical and thermo-
dynamic problems concerned with the production of the gas.
In order to calculate efficiencies we simply assume that a
certain mass of a perfect gas is present. These assumptions
which seem rather reasonable for the purpose of a preliminary
attack, should yield the ideal mechanical performance data
for devices of this type.

4

We outline the contents of this report section by section together with the main conclusions reached in each:

(Section 2. Contents discussed above.)

Section 3. Formulation of the hydraulic theory. The equation of motion of the water during a single stroke is derived from the relation for the energy balance. The formula for the impulse is also obtained. (This is essentially the same as the theory derived in the previously quoted report of F. Zwicky.) The theory leads to an initial value problem for a non-linear ordinary differential equation of second order.

Section 4. Introduction of dimensionless variables and parameters. The essential variables and parameters are expressed in dimensionless form. In all cases involving the determination of quantities as functions of the time through actual integration of the equations of motion (as in part of Section 7 and in Sections 9 to 12 incl.) these variables are very convenient.

Section 5. Duct of variable cross-section: Approximate formulas. Shape of duct for maximum impulse. Although the hydraulic theory as developed in Section 3 is quite elementary from a mathematical point of view, the amount of calculation necessary for a complete discussion of the motion of the water, the impulse I_s per stroke, the ratio of impulse per stroke to mass of fuel injected per stroke (the "effective velocity" or "specific impulse" I_s/M_g), the time of stroke t_s, and the average thrust $\overline{F}_s = I_s/t_s$ in their dependence on all of the essential parameters would be truly enormous. The reason for this is that the equation of motion can not be integrated explicitly, so that numerical integration methods must be used. However, it turns out that the quantities just now enumerated above, which are the most important ones for

practical purposes, can be expressed by approximate
formulas which do not involve a complete integration of the
equation of motion. Section 5 contains a derivation of these
formulas, together with proofs that the formulas yield
rigorous upper bounds for I_s and $\bar{F}_s$ and a rigorous lower
bound for t_s.

By comparison with the results obtained through
accurate numerical integration of the equation of motion
in a large number of cases the formulas are known to furnish
results which are accurate within 2 per cent for I_s and
5-15 per cent for t_s as long as the initial pressure ratio
is greater than 10 and for ducts which do not depart in
shape too much from that of the straight tube of uniform
cross-section.

Section 5 also contains a discussion of the question
of maximizing the impulse with respect to changes in shape
of the hydropulse tube. In case the forward velocity of
the hydropulse is small, it is not difficult to show that
the straight cylindrical tube is as good as any other for
this purpose. This conclusion follows from the fact that
our approximate formula for I_s yields an upper bound for all
shapes of tubes and the fact (determined by actual calculation) that the upper bound is attained within a per cent or
two in the case of the straight tube. However, if the
forward speed of the hydropulse is high, so that the velocity
of the water in the tube is high at the beginning of the
stroke, the impulse per stroke I_s is affected somewhat by
changes in the shape of the tube. We have not obtained a
final result in these cases, but some conclusions from the
present section, combined with a few numerical results
from Section 8 on tubes with conical sections, indicate that
the best shape of tube for maximum impulse during the power
stroke at a given forward speed would be one with a flared-out

forward part joined to a straight cylinder extending to the
exit, as indicated in the accompanying sketch.

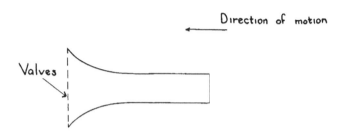

Direction of motion

Valves

<u>Section 6. The tube of constant cross-section: Formulas</u>
<u>showing dependence of essential quantities on all parameters.</u>
The main bulk of our calculations concern the tube of constant
cross-section. Our objective is a complete discussion of the
behavior of all important quantities when any or all parameters
are varied. We have already stated that approximate formulas
are available for tubes of any shape. These formulas are
developed explicitly in Section 6 for the special case of
the straight tube in such a way as to make <u>a complete dis-</u>
<u>cussion possible for the full ranges of all essential</u>
<u>parameters.</u> In particular, formulas for the impulse I_s,
the effective velocity I_s/M_g, (which is a measure of the
efficiency of the device), the time of stroke t_s, and the
average thrust $\bar{F}_s = I_s/t_s$ are obtained. These formulas
exhibit explicitly the dependence on the following parameters:
1) the initial pressure ratio p_1/p_e, 2) the ratio M_g/M_w of
the mass of gas M_g injected per stroke to the initial mass
M_w of water in the tube, 3) the initial velocity U_1 of the
water in the tube,* 4) the adiabatic constant γ , 5) the

---

* The initial velocity U_1 of the water in the tube would be
approximately equal to the forward velocity of the hydropulse.

initial temperature and molecular weight of the gas.

At the end of Section 6 graphs and a table are given which compare the results obtained by the approximate formulas with those obtained by accurate numerical integration of the equation of motion. The approximate formulas are found to be amply accurate for our purposes.

Section 7. The tube of constant cross-section. Performance curves, with discussion. This section is perhaps the most important one in the report. The behavior of the essential quantities with respect to variations of the parameters is given in the form of graphs placed at the end of the section. A brief summary of the main results follows:

a) The effective velocity I_s/M_g, or ratio of impulse per stroke to mass of fuel injected per stroke. The effective velocity increases quite slowly once the initial pressure ratio p_i/p_e becomes greater than about 30, particularly for the higher initial velocities. Thus there would be no great advantage from the point of view of fuel conservation in going to pressure ratios much above 30. The effective velocity decreases steadily to zero as M_g increases steadily to infinity. On the other hand, as M_g tends to zero, the effective velocity increases steadily and tends to a definite limit which is finite if the initial velocity U_1 is not zero, but infinite if U_1 is zero. It is therefore clear that efficient operation is promoted by keeping the amount of fuel injected per stroke as low as that which is compatible with the attainment of other objectives. It is clear that M_g must be large enough so that a sufficiently high thrust can be attained. As the initial velocity U_1 increases, I_s/M_g decreases and has the limit zero. This is to be expected, since the thrust augmentation is lowered with increase in the initial velocity of the water.

The actual numerical values for the effective velocity

in practically reasonable cases are of the order of several hundred thousand feet per second. These are high values compared with those achieved with jet devices in air, the difference being due to the very great density of water as compared with air.

   b) The time of stroke t_s: The time of stroke is not very sensitive to changes in the initial pressure ratio p_1/p_e if this ratio exceeds 30 or 40. When the fuel ratio M_g/M_w alone is varied the following behavior of t_s is found: As M_g/M_w tends to zero so also does the time of stroke t_s, while t_s tends to infinity like $\sqrt{\dfrac{M_g}{M_w}}$ when M_g/M_w tends to infinity. That t_s should tend to zero with M_g seems at first sight rather strange, but it is readily explained by the fact that the initial pressure ratio is assumed to be held fixed while M_g tends to zero. This would be difficult to achieve in actual practice.

   c) The average thrust $\bar{F}_s = I_s/t_s$. The average thrust increases steadily with increase in the initial pressure ratio, as we would expect. When the fuel ratio M_g/M_w alone is varied, the average thrust $\bar{F}_s$ behaves as follows: $\bar{F}_s$ decreases steadily from a finite value for $\dfrac{M_g}{M_w} = \infty$ and tends as $M_g/M_w \rightarrow 0$ to another finite value which is half the value for $M_g/M_w = \infty$. Again this somewhat paradoxical behavior of F_s as the quantity of fuel M_g injected per stroke approaches zero is explained by the fact that the initial pressure ratio is assumed to be maintained as $M_g \rightarrow 0$, and this could not be achieved in practice. Nevertheless, in view of the fact that the efficiency as measured by the effective velocity I_s/M_g has its maximum for $M_g/M_w = 0$, the following important conclusion is strongly indicated: The hydropulse should be operated at high frequency with injection of small amounts of fuel per stroke. The quantity of fuel injected per stroke should be kept as low as is compatible with maintenance of a minimum initial pressure ratio (of the order of 30 or 40 if possible) so that the average thrust will be maintained at the value necessary to overcome the drag on the hydropulse.

  d) Gas pressure and exit velocity of the water as functions of the time. In this part of Section 7 we report a few of our calculations on the behavior of the gas pressure and velocity of the water as functions of the time. These results are of interest in themselves. They also lead to the development of the approximate formulas. The principal observation to be made is that the gas pressure decreases very rapidly initially, while the velocity of the water increases very rapidly at the beginning. Both quantities then remain nearly constant after the first short interval of time at the beginning of the stroke. A number of graphs exhibiting this behavior in a few typical cases are given at the end of Section 7. In Sections 9 and 10 two possible means of smoothing out these initial abrupt variations in pressure and in velocity of the water are discussed.

  e) Effect of linear buildup (in time) of the gas pressure. In all but this portion of Section 7 we assume that the initial high pressure of the gas is created instantaneously. Actually the gas pressure would probably be built up gradually to its maximum after injection of the fuel. In what manner the pressure would rise in actual practice is not known -- it would depend on the chemical reactions and the thermodynamics of the process. In this part of Section 7 we give the results of a few calculations assuming the gas pressure to rise linearly in time to its maximum, after which the gas expands adiabatically down to exit pressure (as in all of our other calculations). For the case of zero initial velocity of the water in the tube curves are given from which

the gas volume and rate of expansion are determined for any
given rate of increase in pressure. The time required for
the pressure to reach its maximum was taken as about 30 per
cent of the entire stroke in one numerical case and 40 per
cent in another. The time of stroke and impulse per stroke
were computed in a number of cases and the results compared
with cases in which the gas pressure reaches its maximum
instantaneously. The results, though inconclusive, are
interesting. In these few cases we find a gradual buildup
in pressure to be favorable for the operation of the hydropulse, since the impulse delivered per stroke for a given
quantity of fuel appears to be substantially larger than in
the case of instantaneous rise of pressure. (However, our
method of comparing the quantities of fuel may be open to
serious objection on physical grounds. For details, see
Section 7e).

If the last conclusion were to be found generally
valid, it would mean that our efficiencies (which were all
computed on the assumption of instantaneous pressure rise)
are on the conservative side.

Section 8. Tubes with conical portions. The results of
calculations giving the impulse and time of stroke for three
different types of tubes having conical portions in front
are presented and compared with those for the straight
cylindrical tube. The purpose and the main results of this
section have already been stated in the course of our remarks
above referring to Section 5.

Section 9. Straight tube with mass and spring for elastic
energy storage. One of the results of the calculations of
Section 7 is that the gas pressure (and consequently the
thrust) and the velocity of the water in the tube change very
rapidly at the beginning of the exhaust stroke. In order to
promote efficiency it seems likely that a more even distribu-

tion of the thrust and of the exit velocity of the water
would be desirable. One possible way to do this would be to
provide for an elastic storage of energy during the early
part of the stroke which would then be delivered again in the
later part of the stroke. Just what means should be taken
to accomplish this in practice is far from clear. What we
have done is to investigate a mechanical model in order to
find out if possible what can be accomplished by such means.
The model is obtained by assuming that the left end of the
exhaust chamber is closed off by a piston of a certain mass
which is attached to a spring (instead of a rigid wall at the
left end.) The calculations were carried out for various
values of the mass. It turns out that the time distribution
of the thrust and the velocity of the water can be very
considerably influenced through proper choice of a mass and
spring. In particular, the abrupt initial variations in these
quantities can be smoothed out rather successfully. (It
might be noted that a spring acting alone without a mass is
not effective.)

It is found that the operation of the hydropulse might
be rather erratic for certain values of the parameters since
the time of stroke may be discontinuous for special values
of the parameters.

Section 10. Spring effect obtained from a layer of gas. One
method of approach to the problem of providing elastic
storage of energy during part of the stroke might be to inject
the fuel in such a way as to create two gas volumes separated
by a mass of water. Various cases of this kind are treated
in Section 10. The number of parameters which could be varied
in these cases is rather large, and our calculations cover
only a small number of the possibilities. However, they do
show that such a method could be used to modify to some extent
the time distribution of thrust and velocity.

Section 11. The valveless hydropulse. Except in this and
the section to follow we assume always that the left end of
the hydropulse tube is closed by valves of some kind. In
Section 11 the possibility of operating a hydropulse without
valves is considered. The tube is assumed to be open in
front but to have a constricted front portion. At the instant
of injection of the gas at relatively high pressure (behind
the constriction, of course) water is assumed to be flowing
through the tube. The motion of the water in the tube is
analyzed, on the basis of our usual assumptions, up to the
time when the gas pressure has decreased to the value at the
exit. The solution of the problem for any one set of
parameter values requires the numerical integration[*] of a
pair of second order non-linear differential equations.
Four cases were treated. The general conclusion reached is
the following: As far as the total impulse for a single stroke
is concerned, a hydropulse without valves could be designed
so that its performance would compare favorably with that of
the hydropulse with valves. To do this, however, a fairly
long narrow forward or nose section would seem to be needed.
(In our calculations we assumed the cross-section area of the
nose to be one-fourth that of the main part of the hydropulse
tube.) As we shall see shortly, such a constricted forward
portion of the tube would be very unfavorable for the
scavenging or refill portion of the whole cycle of operations.

Section 12. Scavenging. We quote from Section 12: "In this
report we have in general concentrated our attention on a
single exhaust stroke of the hydropulse and have left aside
problems concerning other phases of the full cycle, in particular, the phase of 'scavenging'. By this is meant the
portion of the cycle during which the tube is being refilled

---

[*] The method used is explained in Appendix I.

with water in preparation for the injection of fuel and the beginning of a new exhaust stroke." -- "The problem we wish to consider is the following: We assume that the exhaust stroke has just been completed, i.e. that the gas has expanded (with the valves closed) until its pressure has decreased to the value p_e on the outside of the tube and the water in the exhaust chamber has acquired a certain velocity. An instant later the gas will be at a pressure less than p_e (since the inertia of the moving mass of water will lead to a further expansion of the gas,) so that valves could be opened automatically by the pressure of the water in the forward part of the tube.[*] Our purpose in this section is to study the subsequent motion of both layers of water, on the basis of our usual assumptions regarding the behavior of gas and water layers."

Again the solution of the problem requires the numerical integration of a pair of second order differential equations involving a considerable number of parameters. We have carried out the calculations (by the methods of Appendix I) in one case, using what seem to be reasonable values for the parameters. It turns out that the scavenging operation is, generally speaking, very slow. We find in our numerical case that the time required for the gas pressure to reach its minimum after opening the valves is three times as long as the time for the previous power stroke, and even then the refilled volume would be less than that of the gas layer at the instant of opening the valves.

For efficient operation of the hydropulse, it is clear that the scavenging phase should not take up too large a

---

[*] This implies, of course, that the tube extends a certain distance in front of the valves. The tube is also taken to be a cylinder of uniform section throughout.

fraction of the full cycle. Our calculations, although they cover only one case, thus indicate that the scavenging phase in the operation of the hydropulse may present problems of crucial importance. The latter part of Section 12 discusses a few possibilities for overcoming the difficulties, which in their turn lead to new problems (more careful consideration of the actual flow into the mouth of the tube, for example) which will require investigation.

Section 13. Effect of finely distributed gas bubbles. In practice, the fuel is likely to be injected into the water in such a way as to result in the creation of a large number of minute bubbles distributed over a certain volume of the water, rather than in one single layer, as was assumed in earlier sections. The motion of the water can also be treated without difficulty in this case. This problem is treated in Section 13 on the basis of a number of simplifying assumptions. The main conclusion reached is that the motion of the water in the tube is the same in this case as it would be for the corresponding case of a single layer of gas in a tube of smaller total volume. In fact, the volume of the tube with one layer of gas should be diminished by one-third of the volume of that part of the water which contains the gas bubbles in order to make the two cases identical with regard to total impulse delivered per stroke, etc. In other words, only two-thirds of the volume of the water containing the gas bubbles is effective for thrust augmentation.

Appendix I. The results of this report entailed a great deal of numerical computation of various kinds about which nothing much is said in the main part of the report -- to have done so in any detail would have at least doubled the length of an already overlong report. However, a few remarks on numerical methods of two different kinds have been placed in two appendices. Appendix I explains the method of finite differ-

ences which was used for the numerical integrations of the
systems of differential equations encountered in Sections 9
to 12 inclusive. This method is not at all a new or unknown
one, but it was felt that the future usefulness of this
report might be enhanced by the inclusion of a method which
we found convenient and efficient for solving this type of
problem. Future developments might well require refinements
in the treatment of the hydropulse which could lead to still
more complicated systems of differential equations than those
in Sections 9 to 12. However, the method explained in
Appendix I is capable of dealing with such cases by general-
ization in a rather obvious manner.

Appendix II. For dealing with simpler cases in which only
one differential equation occurs the method of finite
differences can be replaced by a method which gives accurate
results with very much less labor in computation. This
method is explained in Appendix II for the special case of
the straight tube. It is, however, equally useful for tubes
of varying cross-section, and this may be of some use in the
future.

Section 2. Assumptions leading to the hydraulic theory.

We make certain simplifying assumptions concerning
the motion of the water in a tube of circular cross-section.
The area of the cross-section may vary along the tube, and

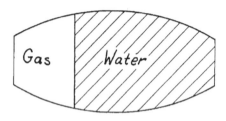

Figure 2.1

the water is assumed to be in motion under the pressure of
a volume of gas, as indicated. The following assumptions are
made:

1. The gas is at every instant in a uniform state
throughout its volume, and the interface between gas and water
is a plane normal to the axis of the tube. The gas is assumed
to expand adiabatically.

2. The water is considered to be incompressible and
non-viscous, and its motion to be irrotational.

3. The hydraulic assumption is made, i.e. that the
motion can be described with sufficient accuracy in terms of
the average velocity over each cross-section.

4. The pressure at the exit is a constant, independent
of the exit velocity of the water.

The first three assumptions seem quite reasonable, at
any rate for the purpose of an investigation having a prelimi-

nary character. Various refinements could and will be made
if a need for them should arise later. The fourth assumption
is a more doubtful one and may prove to be a rather poor
approximation in the cases of practical importance. We have
nevertheless made all of our calculations on the basis of this
assumption. It seems reasonable to suppose that we can at
least make valid comparisons of different cases since it is
to be expected that the inaccuracies arising from the use of
assumption (4) affect the various cases we consider in the
same general way.

On the basis of these assumptions it is quite easy to
obtain a mathematical formulation of the problem of
determining the motion of the water, the thrust exerted on
the tube, and any other relevant quantities.

Section 3. Formulation of the hydraulic theory.

In this section we derive the equation of motion of the water in the tube based on the assumptions of the preceding section. The important fundamental formulas for the thrust, impulse, and time of stroke are also derived.

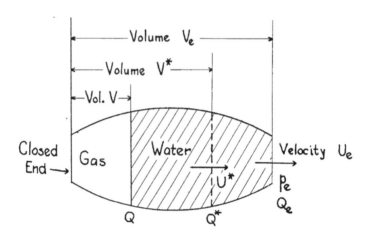

Figure 3.1

We shall assume that the tube moves with constant velocity through the surrounding water. The equations which follow are written with respect to a coordinate system moving with the tube; the coordinates thus form an inertial system.

Some of the quantities which interest us are indicated in Figure 3.1. By Q and U we mean the (circular) cross-section area of the tube and the velocity of the water at any section of the tube. It turns out to be convenient for our purposes to consider that any particular section of the tube is located by giving the volume V^* of the tube between that section and the closed front end of the tube (instead of giving the distance between these sections, for example.) Thus Q is a function of V^*: $Q^* = Q(V^*)$. Since the motion of the water in the tube is (in general) not steady, the velocity U will be a function of the time t as well as of the volume V^*: that is, $U^* = U(V^*, t)$.

The motion of the water results from the pressure exerted by a layer of gas located at the closed end of the tube. We denote by V the volume of the gas at any instant, by P its pressure, and by $\dot{V}$ its time rate of change. These three quantities are thus functions of the time t.[#] However, in all of our work we consider only time intervals during which the gas volume is expanding (that is, motions such that the water is always moving to the right and out of the tube) and hence the time t can be considered a uniquely defined function of V. For the derivation of our formulas it is convenient to consider the pressure P and rate of expansion $\dot{V}$ as being functions of V. The quantity $U^* = U(V^*, t)$, the velocity of the water at any section, may also be expressed in terms of V instead of t: $U^* = U(V^*, V)$.

We make use of subscripts to denote special sections of the tube. The subscript e denotes the exit of the tube, so that p_e and U_e represent the pressure and velocity there, while V_e is the total volume of the tube (or the value of V^* at the exit cross-section, though we dispense with the $*$ in this case[##]). The initial gas pressure, gas volume, and rate of change of gas volume (i.e. their values at the time $t = 0$) are denoted by p_i, V_1, and $\dot{V}_1$ respectively. In all cases we

---

[#] It should be emphasized that these quantities are not functions of V^*.

[##] Later on we are concerned only with the volume V of the gas and can do without the quantity V^* entirely. It is introduced in this section as a necessary aid in deriving the fundamental equations.

determine the motion up to the time when the gas has
expanded down to the "atmospheric" pressure p_e. The gas
volume and its time rate of change at this moment are denoted
by V_s and $\dot{V}_s$ - the symbol p_s for the gas pressure at this
instant is redundant since $p_s = p_e$. The subscript s refers
to the "stroke", which is assumed to end when the gas has
expanded down to the pressure p_e.

We proceed to derive the fundamental equations. It
is convenient to base them on the law of conservation of
energy. This law expresses the equality of the net work done
by the external forces on the original mass of water in the
tube to the change in kinetic energy of the same mass of
water when the gas in the tube expands from the initial volume
V_1 to some volume V. The net work done is then clearly given
by $\int_{V_1}^{V} (p-p_e)dV$, which represents the work done by the expanding gas diminished by the work done in forcing the water out
of the tube against the constant exit pressure p_e. (Here we
make our first use of the assumption that the water is incompressible.) The change in kinetic energy is obtained by
subtracting the kinetic energy K_3 of the water in the tube
at the initial instant (i.e. when the gas volume is V_1) from
the sum of the total energy flux K_1 out of the exit section
(during the expansion from V_1 to V) and the energy K_2 of the
water remaining in the tube at the moment when the gas has
expanded to the volume V.

The basis for the calculation of the kinetic energies
is the following formula for the velocity $U(V^*,V)$ at any
section of area $Q(V^*)$:

(3.1) $$U(V^*,V) = \dot{V}(V)/Q(V^*),$$

which is an immediate consequence of the assumptions (2) and

(3) (see Section 2 above), referring to the incompressibility of the water and the use of the average velocity over any section.[*] In particular, the exit velocity U_e is given by $\dot{V}(V)/Q_e$. When the gas volume V changes by an amount dV, a mass of water ρdV is forced out of the exit section, ρ being the density of the water. It follows that the total energy flux K_1 out of the tube is given by

$$(3.2) \qquad K_1 = \frac{1}{2}\rho \cdot \frac{1}{Q_e^2} \int_{V_1}^{V} \dot{V}^2 dV \ .$$

The kinetic energy K_2 of the water remaining in the tube at the final instant is obtained by integrating $\frac{1}{2}\rho U^2(V^*,V)$ with respect to V^* from V to V_e, V being held fixed; the result is, in view of (3.1):

$$(3.3) \qquad K_2 = \frac{1}{2}\rho \dot{V}^2(V) \int_{V}^{V_e} \frac{dV^*}{Q^{*2}} \ .$$

The kinetic energy K_3 of the water initially in the tube is computed in the same way as for K_2, with $\dot{V}$ taking the special value $\dot{V}_1$ and the limits on the integral extending from V_1 to V_e:

$$(3.4) \qquad K_3 = \frac{1}{2}\rho \dot{V}_1^2 \int_{V_1}^{V_e} \frac{dV^*}{Q^{*2}} \ .$$

---

[*] From (3.1) we observe that the motion of the water is known as soon as the gas volume V is known as a function of the time t.

The equation expressing the energy balance is therefore:

$$(3.5) \quad \int_{V_1}^{V} (p-p_e)dV = \frac{1}{2}\rho\left[\frac{1}{Q_e^2}\int_{V_1}^{V}\dot{V}^2 dV + \dot{V}^2 \cdot \int_{V}^{V_e}\frac{dV^*}{Q^{*2}} - \dot{V}_1^2\int_{V_1}^{V_e}\frac{dV^*}{Q^{*2}}\right]$$

The differential equation of motion can now be obtained from (3.5) by differentiation with respect to V. In doing so, it should be noted that the second term in the brackets in (3.5) furnishes two terms upon differentiation: one resulting from the factor $\dot{V}^2$, the other from the fact that the lower limit of the integral is V. The third term in the bracket is of course independent of V. The equation of motion is

$$(3.6) \quad \frac{p-p_e}{\rho} + \frac{1}{2}\dot{V}^2\left(\frac{1}{Q^2(V)} - \frac{1}{Q_e^2}\right) - \frac{1}{2}\frac{d}{dV}(\dot{V}^2)\cdot\int_{V}^{V_e}\frac{dV^*}{Q^{*2}} = 0,$$

in which the quantity $Q(V)$ in the parenthesis refers to the cross-section area at the gas-water interface. The equation (3.6) can also be derived from the Bernoulli law. The differential equation (3.6) is a first order differential equation for the quantity $\dot{V}$, that is, for the rate of expansion of the gas. If the time were to be reintroduced by the relation $dV = \dot{V}dt$, the result is, of course, a second order differential equation for $V(t)$. In fact, $\frac{1}{2}\frac{d}{dV}(\dot{V}^2) = \frac{1}{2}\frac{d}{dt}(\dot{V}^2)/\frac{dV}{dt} = \ddot{V}$.

So far we have not specified the pressure p of the gas. Since the gas is assumed to expand adiabatically (assumption (1)) we can express p as a function of V by the relation

$$(3.7) \quad pV^\gamma = p_1 V_1^\gamma,$$

in which γ is the adiabatic exponent.

The equations (3.6) and (3.7) together with the values V_1 and $\dot{V}_1$ for the gas volume and its rate of change at the initial instant constitute a complete mathematical formulation of the problem of determining V as a function of t once the shape of the tube and values for the exit pressure p_e and the adiabatic constant γ have been given. The succeeding sections of this report are concerned with the solution of this initial value problem and its interpretation for practical purposes. In the remainder of this section we obtain formulas for various important quantities, which can be computed once the differential equation has been solved for V as a function of t.

Two of the most important basic quantities of interest on practical grounds are the impulse and the thrust made available during the stroke for propulsive purposes. In the above derivations we have fixed our attention on the forces exerted on the <u>water in the tube</u> rather than on the forces exerted on the tube itself. Since we wish to continue our analysis from this point of view because of its convenience, it is necessary to indicate the relation between the net force and the net impulse imparted to the water and the net forward thrust and impulse imparted to the tube. In all that follows the quantities F and I represent the <u>resultant force on the water in the tube</u> at <u>any time</u> and the <u>total impulse delivered to the water in the tube</u> from the initial instant up to any time. Both of these quantities are positive when directed "backwards", i.e. in the direction toward the exit of the tube.

The following Figure 3.2 indicates the relation between F and the forces on the tube.

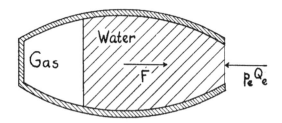

(a) F = total forces on water

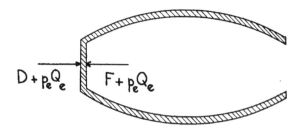

(b) D = total drag

Figure 3.2

The force F on the water arises from the forces exerted by
the tube, the gas, and the water pressure p_e at the exit, the
latter being directed opposite to F. Consequently the tube
exerts a backward force $F + p_e Q_e$ on the water contained in it.
The total force on the tube arising from the water inside it
is thus directed forward and has the magnitude $F + p_e Q_e$. If
the average pressure of the water on the outside of the tube

is greater than p_e, the excess of the average pressure above
p_e will create a drag which we include in the total drag D
(see Figure 3.2b) along with the other components of the
resistance arising from the hydropulse and the object it
propels. Consequently we see that the quantity F we have
defined represents numerically the thrust available to overcome the drag.[*] In the same way the quantity I (which is of
course the time integral of F) is seen to represent the
net impulse available for propulsive purposes.

We obtain first a formula for the impulse I, regarded
as a function of the gas volume V. By the law of conservation
of momentum, it is equal to the momentum of the water remaining
in the tube at the final instant, plus the momentum trans-
ported through the exit, minus the momentum in the water at
the initial instant. By the same considerations as were used
in deriving (3.5) we find

$$(3.8) \quad I(V) = \rho \left[\frac{1}{Q_e} \int_{V_1}^{V} \dot{V} dV + \dot{V} \int_{V}^{V_e} \frac{dV^*}{Q^*} - \dot{V}_1 \int_{V_1}^{V_e} \frac{dV^*}{Q^*} \right].$$

It has already been implied that the time interval t_s
for the complete exhaust stroke is chosen as that extending
from the initial instant t = 0 to the instant when the gas
volume is V_s (or the gas pressure is p_e.) The total impulse
per stroke I_s, which is a fundamentally important quantity
for us, is obtained from (3.8) by setting $V = V_s$. That is:

$$(3.9) \qquad I_s = I(V_s).$$

---

[*] Since we assume the tube to be moving with constant
velocity we have F = D.

The thrust F can be obtained by differentiating the impulse I with respect to the time. We observe that $\frac{dV}{dt} = \dot{V}$ and that $\frac{d^2V}{dt^2} = \ddot{V} = \frac{d\dot{V}}{dt} = \dot{V}\frac{d\dot{V}}{dV}$ and obtain from (3.8) the following formula for the thrust:

$$(3.10) \qquad F = \rho \dot{V}^2(\frac{1}{Q_e} - \frac{1}{Q(V)}) + \rho \dot{V}\,\frac{d\dot{V}}{dV}\int_V^{V_e}\frac{dv^*}{Q^*} \; .$$

It is of great interest to compare the relative quantities of fuel required for the exhaust stroke under various operating conditions. This should be done by considering in detail the reaction of the fuel with the water, but since we have little or no information on this point we have made our calculations referring to fuel consumption assuming that initially we have in each case a layer of an ideal gas of a specified molecular weight at a definite temperature. Thus we fix the mass M_g of fuel used in any given case by the formula:

$$(3.11) \qquad M_g = \frac{\mu}{R\theta_1}\,p_1 V_1 \; ,$$

in which μ is the molecular weight, R is the gas constant per mol, θ_1 is the initial absolute temperature, and p_1 and V_1 are the initial pressure and volume of the gas layer.

One of our principal objects is to study the dependence of the impulse per stroke per unit mass of fuel, that is, the **quantity** I_s/Mg, on variations in all of the relevant parameters. The quantity I_s/M_g has the dimensions of velocity and is called the effective velocity (or specific impulse.) If we set $I_s/M_g = \frac{I_s/T}{M_g/T}$ with T the time for a complete cycle, then we see that the effective velocity represents the ratio of

average thrust to rate of fuel consumption (provided we neglect
the losses in impulse during the refill and scavenging portions
of the full cycle.) Thus the effective velocity is a
reasonable measure of the efficiency of the hydropulse. We
shall study its dependence on variations in initial gas
pressure, mass of fuel, initial velocity of the water in the
tube, and on variations in the adiabatic exponent γ. We
shall also study the dependence of the <u>time of stroke</u> t_s,
and the <u>average thrust over the stroke</u> $\overline{F}_s = I_s/t_s$ on
variations in the same parameters.

  The time of stroke t_s is obtained from the formula

(3.12)
$$t_s = \int_{V_1}^{V_s} \frac{dV}{\dot{V}} ,$$

in which $\dot{V}$ is to be determined by integration of the
equation of motion (3.6).

Section 4. Dimensionless variables and parameters.

For certain of the later sections it is convenient to make use of dimensionless variables and parameters, particularly when the actual integration of the equation of motion (3.6) is in question.

For the dimensionless time τ we write

$$(4.1) \qquad\qquad t = T \cdot \tau \ ,$$

in which the unit of time T is defined by

$$(4.2) \qquad\qquad T = \frac{v_e}{Q_e \sqrt{\frac{2p_e}{\rho}}} \ .$$

All quantities in (4.2) have been defined in the preceding section. The dimensionless gas volume v is defined by

$$(4.3) \qquad\qquad V = V_e \cdot v \ ,$$

while the dimensionless rate of expansion v' is given by the relation

$$(4.4) \qquad\qquad \dot{V} = Q_e \sqrt{\frac{2p_e}{\rho}} \, v' .$$

The prime denotes differentiation with respect to τ. We observe that the quantity $\sqrt{\frac{2p_e}{\rho}}$ has the dimensions of velocity. It is perhaps of interest to note that the value of this unit of velocity is about 32 mi./hr. when p_e is atmospheric pressure.[*]

---

[*] In certain later sections a different unit of velocity is sometimes used, but this occurs only in formulas not written in terms of the dimensionless quantities defined here.

We denote by q the dimensionless cross-section area defined by

(4.5) $$Q = Q_e \cdot q \; .$$

Finally, we introduce the dimensionless quantities u, f, and j corresponding to the velocity of the water U, the thrust F, and the impulse I:

(4.6) $$U = \sqrt{\frac{2p_e}{\rho}} \cdot u \; ,$$

(4.7) $$F = 2p_e Q_e \cdot f \; ,$$

(4.8) $$I = \rho V_e \sqrt{\frac{2p_e}{\rho}} \cdot j \; .$$

In terms of our dimensionless quantities the differential equation of motion (3.6) derived in the preceding section takes the form

(4.9) $$v_s^{\gamma} v^{-\gamma} - 1 + (v')^2 (\frac{1}{q^2} - 1) - \frac{d}{dv}(v')^2 \int_v^1 \frac{dv}{q^2} = 0,$$

in which $v_s = V_s/V_e$ and the prime refers to differentiation with respect to τ .

Section 5. Duct of variable cross-section. Approximate formulas. Shape of duct for maximum impulse.

The integration of the differential equation of motion (3.6) is not difficult, but very laborious, since it must be done numerically. Consequently a complete discussion of the solutions of the differential equation and of the important quantities, such as the total impulse and the time of stroke, in their dependence on all of the many parameters would involve a mountainous labor in calculation if it were necessary to carry out numerical integrations in every case. Fortunately, it turns out that all of the more important quantities can be given by relatively simple and sufficiently accurate approximate formulas, which contain all of the parameters explicitly and thus lend themselves well to a complete discussion. These formulas have been checked by comparison with the results obtained by numerical integration of the differential equation in quite a large number of cases (including two types of tubes with conical front portions, as well as straight tubes) so that limits for the accuracy of the approximate formulas are known (and will be given later.) We proceed to derive the approximate formulas.

The basis of the approximate formulas is the fact, observed in many cases which were computed exactly, that the pressure in the gas decreases very rapidly initially and then stays relatively constant.[*] At the same time, the mass of water in the tube is strongly accelerated initially and quickly reaches a high velocity which then remains nearly constant over the remainder of the stroke. This indicates that the motion of the water may be analyzed with good accuracy on an

---

[*] This results largely because the gas is assumed to obey the adiabatic law so that the pressure decreases rapidly with increase in volume.

impulse basis simply by assuming that the mass of water involved remains constantly equal to the initial mass in the tube.

We denote by E_s the energy imparted to the water as the net result of the work done by the gas pressure and the pressure at the exit of the tube; thus E_s is given by

$$(5.1) \qquad E_s = \int_{V_1}^{V_s} (p - p_e)\, dV ,$$

in which V_1 and V_s represent the initial and final volumes of the gas, V_s being the gas volume at the end of the stroke when the gas pressure has the value p_e. We denote by M_w the initial mass of water in the tube and by E_1 and M_1 the kinetic energy and momentum of the water in the tube at the initial instant. We then use the energy law to express E_s by the following approximate formula:

$$(5.2) \qquad E_s = \frac{1}{2} M_w \bar{U}_s^2 - E_1 ,$$

in which $\bar{U}_s^2$ is the average of the square of the final velocities of the water particles in the tube. The impulse per stroke I_s -- a fundamentally important quantity for us -- can be approximated as follows:[*]

$$(5.3) \qquad I_s = M_w U_s - M_1$$

---

[*] It should be clear that our approximation refers to the quantity I_s as it was defined in Section 3.

in which U_s is the average of the final velocity of the water particles in the tube. We now make the assumption that $\bar{U}_s^2 = U_s^2$, i.e. that the average of the squares of the velocities is equal to the square of the average velocities, (this assumption is exact for the case of a tube of constant cross-section) and eliminate U_s and $\bar{U}_s^2$ from (5.2) and (5.3) to obtain the important approximate formula

$$(5.4) \qquad I_s = \sqrt{2M_w E_s + 2M_w E_1} - M_1$$

for the total impulse per stroke in terms of quantities which can be calculated or estimated in any given case without much difficulty. An important special case is that in which the initial velocity U_1 of the water in the tube is zero, and I_s is given by the simple formula

$$(5.5) \qquad I_s = \sqrt{2M_w E_s} .$$

The formula (5.4) can be interpreted as the result of replacing the hydropulse by a simpler mechanical model: The formula furnishes the total impulse on a rigid body of mass M_w which moves in a straight line under the action of forces which do work of amount E_s. The quantities U_s^2 and $\bar{U}_s^2$ in (5.2) and (5.3) are then both equal to the square of the final velocity of the rigid body.

The time of stroke t_s is determined by the approximate formula

$$(5.6) \qquad t_s = \frac{(V_s - V_1)/Q_e}{U_s} ,$$

in which Q_e is the cross-section area of the tube at the exit, and U_s is to be determined from (5.3). That (5.6) furnishes some sort of approximation to the correct value for t_s is clear.

Finally, the average thrust $\bar{F}_s$ on the water in the tube during the stroke is given approximately by

$$(5.7) \qquad \bar{F}_s = I_s/t_s \ ,$$

with I_s and t_s to be determined by (5.4) and (5.6).

It is important to know that the <u>approximate value</u> given by (5.4) for I_s is a rigorous upper bound for this quantity. We proceed to prove this statement. From the formulas (3.8) and (3.5) of Section 3 we have

$$(5.8) \qquad I_s + M_1 = \rho \left[\frac{1}{Q_e} \int_{V_1}^{V_s} \dot{V} dV + \dot{V}_s \int_{V_s}^{V_e} \frac{dV^*}{Q^*} \right] ,$$

and

$$(5.9) \qquad E_s + E_1 = \frac{1}{2}\rho \left[\frac{1}{Q_e^2} \int_{V_1}^{V_s} \dot{V}^2 \, dV + \dot{V}_s^2 \int_{V_s}^{V_e} \frac{dV^*}{Q^{*2}} \right].$$

We shall write the two integrals in (5.8) as one single integral, first writing the second of the integrals in (5.8) as $\int_{V_s}^{V_e} \frac{dV}{Q}$, which is legitimate since the variable of integration is a dummy variable. Since the intervals of integration in the two integrals (3.3) are contiguous and do not overlap, we may write

$$(5.10) \qquad I_s + M_1 = \rho \int_{V_1}^{V_e} H(V) dV$$

CONFIDENTIAL

where

$$H(V) = \begin{cases} \dot{V}/\mathcal{Q}_e & \text{for} \quad V_1 \le V \le V_s \\[2em] \dot{V}_s/\mathcal{Q} & \text{for} \quad V_s \le V \le V_e \end{cases}$$

Similarly,

(5.11)
$$E_s + E_1 = \frac{1}{2}\rho \int_{V_1}^{V_e} H(V)^2 dV \ .$$

Applying the Schwarz inequality to (5.10), we have

(5.12)
$$I_s + M_1 \le \rho \, (V_e - V_1)^{1/2} \left[\int_{V_1}^{V_e} H(V)^2 dV \right]^{1/2} \ .$$

(The Schwarz inequality asserts that

$$\int f(V)g(V)dV \le \sqrt{\int f(V)^2 dV \cdot \int g(V)^2 dV} \ ,$$

equality holding only if the ratio of $f(V)$ to $g(V)$ is a constant. To apply it to (5.10), take $f(V) = 1$ $g(V) = H(V)$).
      Using (5.11) and the fact that the initial mass M_w of the water in the tube is $M_w = \rho(V_e - V_1)$, we obtain

$$I_s + M_1 \le \sqrt{2M_w(E_s + E_1)}$$

or

(5.13)
$$I_s \le \sqrt{2M_w E_s + 2M_w E_1} - M_1$$

In particular, if the initial velocity of the water in the

tube is zero, we have

(5.14) $$I_s \leq \sqrt{2M_w E_s} \, .$$

Thus, the approximate formulas (5.4) and (5.5) yield rigorous upper bounds for I_s.

It is now possible to draw a number of conclusions of some practical importance regarding the effect of changes in shape of the hydropulse tube on the impulse per stroke. For this purpose it is convenient to write the inequality (5.13) in the form

(5.13)' $$I_s + M_1 \leq \sqrt{2M_w E_s + 2M_w E_1} \, ,$$

and to recall (cf. equation (5.10)) that $I_s + M_1 = \rho \int_{V_1}^{V_e} H(V) dV$,

and that (5.13)' results from application of the Schwarz inequality to the integral. Our calculations for many cases show that the upper bound for $I_s + M_1$ furnished by (5.13)' is practically attained for the straight cylindrical tube (cf. the table at the end of the following Section 6) as well as for a number of tubes having conical portions (see Section 8).[*] The reason for this lies in the behavior of $\dot{V}$ as a function of V over the stroke: Typical cases for the straight tube are shown graphically in Figure 7.9 of Section 7, which shows that $\dot{V}$ increases very rapidly at first but then remains constant -- in other words, $\dot{V}$ is practically constant over most of the range. Hence the function $H(V)$ also behaves like a constant

---

[*] In Sections 6 and 8 the dimensionless variables defined in Section 4 are used rather than the variables of the present section.

(cf. the definition following equation (5.10)) if the cross-
section area of the tube does not vary too much so that
the Schwarz inequality is practically an equality.

Let us consider first cases in which the initial
velocity of the water in the tube is zero, so that the initial
momentum M_1 is zero. In this case formula (5.14) applies,
of course.

It can now be seen that <u>variations in the shape of the
tube in these cases have only a slight influence on the impulse
per stroke</u>, at least for tubes which are not too radically
different from the straight tube: This is an immediate consequence of the fact that the upper bound for I_s given by (5.14)
is practically attained for the straight tube, while on the
other hand the right side of (5.14) is independent of the
shape of the tube -- it depends only on the volume occupied
by the original mass of water M_w in the tube and on the energy
imparted to this mass (which depends only on the initial
pressure, volume, and character of the gas and not on the
shape of the tube.) Thus if M_w is kept fixed the straight
tube will yield very closely the maximum impulse with respect
to variations in shape. The determination of the exact shape
for maximizing the impulse is therefore settled in case the
forward velocity of the hydropulse is not too great.

The cases in which the initial velocity of the water
in the tube is not zero are not so easily dealt with. Of
course, if the initial velocity is not too great (i.e. if the
forward velocity of the hydropulse is not too great) then our
conclusion that the straight tube is as good as any other for
the purpose of maximizing the impulse would still be reasonable. For high velocities, however, the quantity M_1 in (5.13)
appears to be sensitive to changes in shape of the tube. In
Section 8, where a few cases of tubes with conical portions
are treated, the impulse for an initial velocity of about 90

mi./hr is increased by from 10 per cent to 20 per cent by
taking a conical tube in place of a straight cylindrical tube.[*]
For the higher velocities of the hydropulse we have
not settled the question of the best shape of tube for maximizing the impulse, and more work should be done on this
problem. However, we have obtained one further partial result
which is not without interest: Consider only that part of
the tube between V_s and V_e, i.e. that part which is still
filled with water at the end of the stroke. By applying the
Schwarz inequality to the second integral on the right-hand
side of (5.8) it can easily be shown that the impulse is a
maximum when this part of the tube has a constant cross-section,
if the volume of this part is assumed to be fixed in all cases
under comparison.[**] This result, together with the fact that
the calculations of Section 8 show the impulse at high initial
velocities to be increased by flaring out the front section,
indicates that the best shape of tube for maximizing the
impulse is one with a flared out front section attached to a
straight cylindrical portion extending to the exit section.
We proceed next to show that the approximate formula
(5.6) for t_s furnishes a rigorous _lower_ bound for this
quantity in the case of the tube of constant cross-section.
It also undoubtedly furnishes a lower bound for tubes which
do not depart all too radically from the straight cylindrical
tube, but we do not discuss this question here. (See, however,
Section 8 where t_s is calculated for two different types of
conical tubes, which depart rather radically from the straight
shape. The approximate formula yields t_s too low in these

---

[*] Upon what basis the comparison should be made is not
entirely clear. This point is discussed in Section 8.

[**] Without such a restriction the total impulse could easily
be seen to have no finite upper bound.

cases also.) We begin with the exact formula (3.12) for t_s:

$$(5.15) \qquad t_s = \int_{V_1}^{V_s} \frac{dV}{\dot{V}} .$$

We write (5.15) in the following form by adding a certain integral to both sides:

$$(5.16) \quad t_s + \frac{1}{\dot{V}_s} \int_{V_s}^{V_e} \frac{Q^*}{Q_e} \, dV^* = \frac{1}{Q_e} \left[\int_{V_1}^{V_s} \frac{dV}{\dot{V}/Q_e} + \frac{1}{\dot{V}_s} \int_{V_s}^{V_e} Q^* dV^* \right] ,$$

in which $Q^* = Q(V^*)$, as defined in Section 3. Through introduction of the function $H(V)$ defined above (cf. equation (5.10) and the remarks which immediately precede it) and a division of both sides of (5.16) by $V_e - V_1$ we obtain the relation

$$(5.17) \qquad \frac{Q_e}{V_e - V_1} \left[t_s + \frac{1}{\dot{V}_s} \int_{V_s}^{V_e} \frac{Q}{Q_e} \, dV \right] = \frac{1}{V_e - V_1} \int_{V_1}^{V_e} \frac{dV}{H(V)} ,$$

in which the dummy variable V^* has been replaced by V. A known inequality asserts (see the book of Hardy, Littlewood, Polya on Inequalities, p. 151) that if $\phi(y)$ is convex -- i.e. if $\phi''(y) \geq 0$ -- then

$$\frac{1}{b-a} \int_a^b \phi(f(V)) dV \geq \phi \left[\frac{1}{b-a} \int_a^b f(V) dV \right] .$$

Applying this inequality to the right-hand side of (5.17) with $\phi(y) = \frac{1}{y}$ and $f(V) = H(V)$, we obtain:

(5.18) $\quad \dfrac{Q_e}{V_e - V_1} \left[t_s + \dfrac{1}{\dot{V}_s} \int_{V_s}^{V_e} \dfrac{Q}{Q_e} \, dV \right] \geq \dfrac{V_e - V_1}{\int_{V_1}^{V_e} H(V) dV}$.

By making use of (5.10) (an exact formula, it should be noted) to replace $\int_{V_1}^{V_e} H(V) dV$ and noting that $\rho(V_e - V_1) = M_w$ we obtain from (5.18) the result

(5.19) $\quad t_s \geq \dfrac{M_w}{I_s + M_1} \cdot \dfrac{V_e - V_1}{Q_e} - \dfrac{1}{\dot{V}_s} \int_{V_s}^{V_e} \dfrac{Q}{Q_e} \, dV$.

We know from our calculations that the lower bound furnished by the right-hand side of (5.19) is practically attained for all tubes not too radically different from the straight tube. Unfortunately, the right-hand side of (5.19) contains the quantity $\dot{V}_s$ which can be determined only by integration of the equation of motion. However, in the most important case -- that of the tube of constant cross-section -- it is possible to eliminate $\dot{V}_s$, as follows:

We know that $\dot{V}$ is an increasing function of V for $V_1 \leq V \leq V_s$. From (5.10) we may therefore write in this case $(Q(V) \equiv Q_e)$:

(5.20) $\quad I_s + M_1 \leq \rho \dot{V}_s \cdot \dfrac{V_e - V_1}{Q_e}$, or

(5.21) $\quad \dfrac{1}{\dot{V}_s} \leq \dfrac{\rho}{Q_e} \cdot \dfrac{V_e - V_1}{I_s + M_1}$.

Substituting from (5.21) in (5.19) we obtain

$$(5.22) \qquad t_s \geq \frac{M_w}{I_s + M_1} \cdot \frac{V_e - V_1}{Q_e} - \frac{\rho}{Q_e} \cdot \frac{V_e - V_1}{I_s + M_1} \cdot (V_e - V_s)$$

or, since $\rho(V_e - V_1) = M_w$:

$$(5.23) \qquad t_s \geq \frac{M_w}{I_s + M_1} \cdot \frac{V_s - V_1}{Q_e} .$$

Since $(I_s + M_1)/M_w = U_s$ (from (5.3)), we see that the right-hand side of (5.23) is the same as that of (5.6). Hence (5.6) furnishes a rigorous lower bound for t_s in the case of the tube with uniform cross-section.

If I_s and t_s are calculated using (5.4) and (5.6) we may now conclude that the formula (5.7) for the average thrust $\bar{F}_s$ yields a rigorous upper bound for the straight tube, since t_s is a lower bound in this case and I_s is in all cases an upper bound.

The accuracy of the approximate formula (5.6) -- i.e. of (5.23) using the sign of equality -- is discussed at the end of the following Section 6. In this case the formula is found to be more accurate, that is, the lower bound is more nearly attained, when the initial velocity is high rather than low (the opposite was true for the impulse). The approximation is good enough for practical purposes.

Section 6. The tube of constant cross-section. Formulas
showing dependence of essential quantities on all parameters.

In this section the approximate formulas of the
preceding section are adapted to the special case of the tube
of constant cross-section. The integrations are carried out
and the formulas are written in such a way as to make the
discussion of the dependence on the various parameters a
relatively simple matter. The actual discussion is then
carried out in the following Section 7. The essential quantities for which formulas are obtained are: 1) the effective
velocity I_s/M_g 2) the time of stroke t_s, and 3) the
average thrust $\bar{F}_s$ over the stroke. In each of these formulas
the dependence on all of the following parameters is exhibited:
1) the "fuel parameter" M_g/M_w, i.e. the ratio of the mass of
gas per stroke to the mass of water in the tube, 2) the
initial gas pressure ratio p_1/p_e, 3) the initial velocity[*]
U_1 of the water in the tube, 4) the adiabatic exponent γ.

We begin by calculating the quantity E_s in the
formula (5.4). For it we have (from (5.1)):

$$(6.1) \qquad E_s = \int_{V_1}^{V_s} \left(\frac{p_1 V_1^{\gamma}}{V^{\gamma}} - p_e \right) dV \, ,$$

from the adiabatic gas law. The integration can be carried
out to yield

$$(6.2) \qquad E_s = p_1 V_1 \left[\frac{1}{\gamma-1} \left(1 - \left(\frac{V_1}{V_s}\right)^{\gamma-1}\right) - \frac{p_e}{p_1}\left(\frac{V_s}{V_1} - 1\right)\right] \, .$$

---

[*] This velocity might be considered to be approximately
equal to the forward velocity of the hydropulse.

Again using the adiabatic gas law we may write for V_s the equation

$$(6.3) \qquad p_e V_s^{\gamma} = p_1 V_1^{\gamma} .$$

Elimination of V_s from (6.2) yields

$$(6.4) \qquad E_s = p_1 V_1 [\frac{1}{\gamma-1}(1 - \frac{\gamma}{(p_1/p_e)^{1-\frac{1}{\gamma}}} + \frac{1}{p_1/p_e}] .$$

The mass M_g of the gas is related to $p_1 V_1$ by the ideal gas law (3.11); replacing $p_1 V_1$ by means of this relation we obtain finally for E_s the formula

$$(6.5) \qquad E_s = \frac{R\Theta_1}{\mu} \cdot M_g \cdot [\frac{1}{\gamma-1}(1 - \frac{\gamma}{(p_1/p_e)^{1-\frac{1}{\gamma}}}) + \frac{1}{p_1/p_e}] .$$

The formula (5.4) for the impulse per stroke I_s can be written in the form

$$(6.6) \qquad I_s = (I_s)_0 \left\{ \sqrt{1 + \frac{M_w^2 U_1^2}{(I_s)_0^2}} - \frac{M_w U_1}{(I_s)_0} \right\},$$

with

$$(6.7) \qquad (I_s)_0 = \sqrt{2 M_w E_s} .$$

The quantity $(I_s)_0$ represents the impulse per stroke in the special case in which the initial velocity of the water in the tube is zero. For the effective velocity I_s/M_g we may therefore write

$$(6.8) \qquad I_s/M_g = \frac{(I_s)_o}{M_g} \left\{ \sqrt{1 + \left[\frac{\frac{M_w}{M_g} U_1}{\frac{(I_s)_o}{M_g}} \right]^2} - \frac{\frac{M_w}{M_g} U_1}{\frac{(I_s)_o}{M_g}} \right\},$$

with

$$(6.9) \quad (I_s)_o / M_g = \sqrt{2\frac{R\Theta_1}{\mu} \cdot \frac{M_w}{M_g}} \cdot \sqrt{\frac{1}{\gamma-1}\left(1 - \frac{\gamma}{(\frac{p_1}{p_e})^{1-\frac{1}{\gamma}}}\right) + \frac{1}{\frac{p_1}{p_e}}} .$$

If we were to divide equations (6.8) and (6.9) on both sides by $\sqrt{2\frac{R\Theta_1}{\mu}}$, which has the dimensions of a velocity, they would be in dimensionless form. All the essential parameters are exhibited explicity in these formulas.

It is convenient to introduce special symbols for two combinations of quantities in the formula (6.8), as follows: We set

$$(6.10) \qquad h = \sqrt{\frac{1}{\gamma-1}\left(1 - \frac{\gamma}{(\frac{p_1}{p_e})^{1-\frac{1}{\gamma}}}\right) + \frac{1}{\frac{p_1}{p_e}}} , \text{ and}$$

$$(6.11) \qquad k = \sqrt{\frac{M_w}{M_g}} \cdot \frac{U_1}{\sqrt{2\frac{R\Theta_1}{\mu}}} ,$$

and obtain the following relation for the effective velocity, replacing (6.8):

$$(6.12) \qquad \frac{I_s}{M_g} = \sqrt{2\frac{R\Theta_1}{\mu} \cdot \frac{M_w}{M_g}} \cdot \left\{ \sqrt{h^2 + k^2} - k \right\} .$$

The advantage of this form is that h contains (aside from γ) only the pressure ratio p_1/p_e, while k is the only quantity which involves the initial velocity U_1.

We proceed next to amplify the formula (5.6) for the time of stroke t_s: $t_s = \dfrac{(V_s - V_1)/Q_e}{U_s}$ for the case of the tube of uniform cross-section. This is done by following much the same procedure as above, including the introduction of the quantities h and k defined by (6.10) and (6.11). The result for t_s is readily found to be:

$$(6.13) \qquad t_s = \frac{V_e/Q_e}{2p_e \sqrt{\dfrac{R\Theta_1}{2\mu}}} \sqrt{\frac{M_g}{M_w}} \cdot \frac{\left(\dfrac{p_1}{p_e}\right)^{\frac{1}{\gamma}} - 1}{\left(\dfrac{p_1}{p_e}\right) \cdot \sqrt{h^2 + k^2}}$$

Finally, the average thrust $\bar{F}_s$ over the stroke is given by the quotient of I_s and t_s. The result is

$$(6.14) \qquad \bar{F}_s = \frac{I_s}{t_s} = 2p_e Q_e \cdot \frac{\dfrac{p_1}{p_e}}{\left(\dfrac{p_1}{p_e}\right)^{\frac{1}{\gamma}} - 1} \cdot \sqrt{h^2 + k^2}\left(\sqrt{h^2 + k^2} - k\right) \cdot$$

The formulas (6.10) to (6.14) inclusive make an almost complete discussion of the ideal performance characteristics of the hydropulse possible. They do not, of course, make it possible to discuss the motion of the water as a function of time. Also, these formulas are approximations, though rather accurate ones. In fact, the effective velocity is in error by less than 3 per cent if the pressure ratio p_1/p_e is greater than 10. The time of stroke t_s and the average thrust $\bar{F}_s$ are subject to an error of about 5 per cent if $p_1/p_e \geq 10$ and the initial velocity U_1 is greater than about 30 miles per hour; if $U_1 \approx 0$ the error is not greater than 15 per cent.

The curves of Figure 6.1 indicate the accuracy of the approximate formula for the impulse in the case $\gamma = 1.5$, $U_1 = 0$.

The curves give the values of the dimensionless impulse j_s (defined by equation (4.8) of Section 4 for various values of the initial pressure ratio p_1/p_e as functions of the initial volume $v_1 = V_1/V_e$. The points indicated by circles were obtained by numerical integration of the equation of motion and are subject to an error of not more than 1 per cent. The curves, on the other hand, were computed by using (6.8). As we see, the approximate formula furnishes quite accurate results over a wide range of initial pressures and initial gas volumes. The following table gives a number of additional comparisons for j_s when γ and U_1 are varied. (The quantity v_1' is the dimensionless initial velocity defined in Section 4. If p_e is atmospheric pressure, then $v_1' = 1$ corresponds to a velocity of 31.8 miles/hr.)

Accuracy of Approximate Formula

for Impulse

Cases			Impulse j_s	
			Approximate	Exact
$\gamma = 1.5$	$v_1 = .01$	$v_1' = 1$	1.56	1.57
$v_s = 0.5$		$v_1' = 3$	.82	.84
	$v_1 = .05,$	$v_1' = 3$	.069	.068
$\gamma = 1.1$		$v_1' = 0$	1.373	1.366
$v_s = 0.5$				
$v_1 = .01$		$v_1' = 3$	.302	.301
$\gamma = 1.3$		$v_1' = 0$	1.198	1.188
$v_s = .416$				
$v_1 = .024$		$v_1' = 1$	.570	.566

As the table shows, the approximate formula appears to be quite accurate over a wide range of parameter values.

In the following table a few comparisons of exact and approximate values for the time of stroke τ_s (in dimensionless form, cf. Section 4) are given:

Cases			Time of Stroke τ_s	
			Approximate	Exact
$\gamma = 1.1$ $v_i = .01$	$v_i' = 0$		.353	.392
$v_s = 0.5$	$v_i' = 3$		.148	.150
$\gamma = 1.3$ $v_i = .024$	$v_i' = 0$		.319	.370
$v_s = .416$	$v_i' = 1$		.248	.259

As remarked above, the approximate formula for the time of stroke becomes more accurate when the initial velocity (v_i' in dimensionless form) increases.

As we have seen in the preceding section, our formula for I_s gives an upper bound for this quantity. The formula for t_s, as we have seen, furnishes a lower bound, and consequently the formula for F_s yields an upper bound.

The following section is devoted to a discussion of the formulas developed here, largely on the basis of graphs. We also include some results obtained by numerical integration of the differential equation -- in particular, the gas pressure is given as a function of time in a variety of cases.

FIG 6.1   CONFIDENTIAL   47

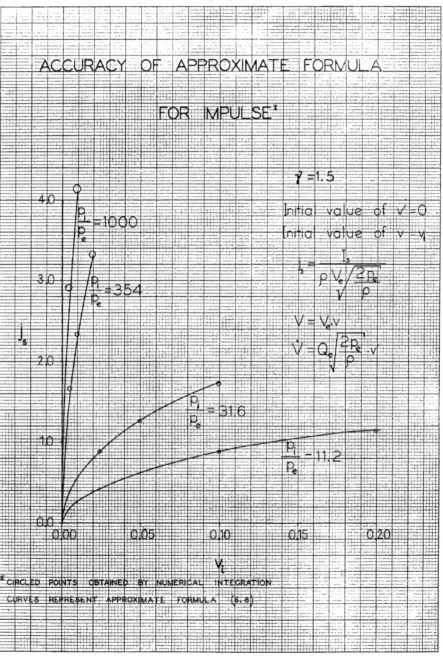

ACCURACY OF APPROXIMATE FORMULA

FOR IMPULSE*

$\gamma = 1.5$

Initial value of $v' = 0$

Initial value of $v = v_i$

$$i_s = \dfrac{I_s}{\rho\, V_e \sqrt{\dfrac{2 p_e}{\rho}}}$$

$V = V_e v$

$$\dot{V} = Q_e \sqrt{\dfrac{2 p_e}{\rho}} \cdot v$$

$\dfrac{p_i}{p_e} = 1000$

$\dfrac{p_i}{p_e} = 354$

$\dfrac{p_i}{p_e} = 31.6$

$\dfrac{p_i}{p_e} = 11.2$

*CIRCLED POINTS OBTAINED BY NUMERICAL INTEGRATION

CURVES REPRESENT APPROXIMATE FORMULA (6.8)

Section 7. The tube of constant cross-section. Performance curves, with discussion.

In Section 7, which in many ways is the heart of this report, we discuss the curves and formulas which furnish the ideal performance characteristics of the hydropulse for the case of a straight cylindrical tube where any of the parameters are varied. This section is broken into five parts each devoted to a special quantity or to a special aspect of our problem, as follows: a) The effective velocity I_s/M_g, b) The time of stroke t_s, c) The average thrust $F_s = I_s/t_s$, d) Discussion of various quantities, such as the gas pressure and exit velocity, as functions of the time, e) Effect of a gradual buildup (in time) of the gas pressure.

a) The effective velocity

The formula for the effective velocity (cf. equation (6.12) of the preceding section) is

$$(7.1) \qquad I_s/M_g = \sqrt{\frac{2R\Theta_1}{\mu} \cdot \frac{M_w}{M_g}} \cdot \left\{ \sqrt{h^2 + k^2} - k \right\},$$

with

$$(7.2) \qquad h = \sqrt{\frac{1}{\gamma - 1}\left[1 - \frac{\gamma}{(\frac{p_1}{p_e})^{1-\frac{1}{\gamma}}} \right] + \frac{1}{\frac{p_1}{p_e}}} \,, \text{ and}$$

$$(7.3) \qquad k = \sqrt{\frac{M_w}{M_g}} \cdot \frac{U_1}{\sqrt{2\frac{R\Theta_1}{\mu}}} \,.$$

For ease in computation of the effective velocity we give three graphs, Figures 7.1, 7.2, 7.3, as follows:

Figure 7.1 is a graph of $\sqrt{2\dfrac{R\Theta_1}{\mu}\dfrac{M_w}{M_g}}$ as a function of the "fuel parameter", i.e. the ratio M_g/M_w of the mass of gas per stroke to the mass of water in the tube initially; both quantities are plotted on a logarithmic scale, for a series of values of the initial temperature Θ_1 ranging from 3000° Kelvin down to 500° K. Figure 7.2 gives the quantity h (cf. (7.2)) as a function of the pressure ratio p_1/p_e for various values of the adiabatic exponent γ. Figure 7.3 shows the quantity $(\sqrt{h^2+k^2}-k)$ (cf. (7.1)) as a function of k for various values of h.

Computation of the effective velocity in any given case is then carried out as follows: Compute first the quantity $\sqrt{2\dfrac{R\Theta_1}{\mu}}$ (dimensionally a velocity) which occurs in several formulas. The value of $\sqrt{2\dfrac{R\Theta_1}{\mu}\dfrac{M_w}{M_g}}$ is then fixed by the graphs of Figure 7.1 once the ratio M_g/M_w is known. The quantity k is than fixed by (7.3) once the initial velocity U_1 is given. For a given pressure ratio p_1/p_e the quantity h is then determined from the graphs of Figure 7.2, after which the quantity $\sqrt{h^2+k^2}-k$ can be determined from the graphs of Figure 7.3. In this way both factors in the formula (7.1) for the effective velocity are determined.

Finally, Figure 7.4 gives graphs showing the dependence of the effective velocity on the fuel parameter M_g/M_w for various initial velocities U_1. These curves are all drawn assuming $h = 1.1$, $\mu = 2$, and $\theta_1 = 2000^\circ$K.

The essential information regarding the dependence of the effective velocity on all parameters is clearly contained in these four sets of curves. Certain pertinent general conclusions should, however, be drawn from them. To begin with, we observe that the quantity h (see Figure 7.2) increases only slightly with increase of the pressure ratio p_1/p_e once

the pressure ratio becomes larger than 30 or 40. Consequently
the same is true of the effective velocity, since it depends
on the pressure ratio only through the quantity h (cf. equations
(7.1) and (7.2)). From this we draw an important practical
conclusion: <u>The efficiency of the hydropulse is not greatly
increased by going to pressure ratios much above 40</u>. We
observe also that the values of h for any reasonable value
of γ and for pressure ratios near 40, say, are close to unity.
Figure 7.4, which is drawn for h = 1.1, thus furnishes graphs
for reasonable pressure ratios. Since the initial velocity
U_1 of the water occurs only in the quantity k in (7.1), it
follows that Figure 7.3 indicates the behavior of the effective
velocity with increase in initial velocity: The effective
velocity decreases steadily with increase in U_1, tending to
zero as U_1 increases indefinitely. This was, of course, to
be expected since the thrust augmentation naturally decreases
as U_1 increases.

Figure 7.4 is one of the most important graphs, since
it gives an indication of the efficiency as a function of the
quantity of fuel used per stroke. The effective velocity
decreases steadily with increase in the fuel parameter, with
its maximum occurring for $M_g/M_w = 0$. The maximum is infinite
when the initial velocity U_1 is zero, but is finite for non-zero
initial velocities and steadily decreases as U_1 increases.
This behavior of the effective velocity comes about, of course,
because the thrust augmenting mass M_w grows large with respect
to M_g. One should also bear in mind that the pressure ratio
p_1/p_e is considered to be kept constant -- it is not at all
obvious that that could be accomplished in practice when M_g
is very small. The practical conclusion to be drawn is this:
<u>The quantity of fuel injected per stroke should be kept as
low as possible, insofar as that can be reconciled with the
attainment of other objectives such as the creation of a certain
minimum average thrust.</u>

b) The time of stroke t_s

The formula for the time of stroke t_s (see (6.13)) is

$$(7.4) \qquad t_s = \frac{V_e/\rho_e}{2p_e\sqrt{2\frac{R\theta_1}{\mu}}} \cdot \sqrt{\frac{M_g}{M_w}} \cdot \frac{(\frac{p_1}{p_e})^{\frac{1}{\gamma}} - 1}{(\frac{p_1}{p_e})\sqrt{h^2 + k^2}} \cdot$$

The quantities h and k are defined in (7.2) and (7.3).

The formula for t_s shows clearly that t_s tends to zero when the fuel parameter M_g/M_w tends to zero (since $k \to \infty$ when $M_g/M_w \to 0$.) At first sight this seems rather paradoxical, since one would be inclined to argue that the time required for an entire stroke would tend to grow very large when the amount of fuel injected is very small. However, it must be remembered that M_g alone is being varied while the initial pressure p1 is kept fixed. This means that the initial volume V_1 is being made smaller, so that a progressively smaller volume of water would be forced out of the tube per stroke, during the expansion of the gas down to the exit pressure p_e. Hence one sees that it is after all not unreasonable that $t_s \to 0$ when $M_g/M_w \to 0$ if p1/p_e is held fixed. On the other hand, when M_g/M_w grows large, it is readily seen that t_s increases essentially as the square root of this quantity, since then k tends to zero.

As an aid in the calculation of t_s from (7.4) we have drawn (see Figure 7.5a and 7.5b) curves for the third factor on the right-hand side of (7.4) as a function of the pressure ratio p1/p_e for various values of k; two such sets of curves are given -- one for $\gamma = 1.4$ and another for $\gamma = 1.2$. From Figure 7.5 we can infer how t_s varies with respect to changes in the pressure ratio and the initial velocity. As we see, t_s decreases as the pressure ratio increases (beyond

53

the value 10, at least) but the decrease in time of stroke is rather slow for pressure ratios above 40. The time of stroke of course decreases as the initial velocity U_1 increases.

   c)   The average thrust $\bar{F}_s = I_s/t_s$.

      The formula for the average thrust, as derived in the preceding section (cf. (6.14)) is

$$(7.5) \quad \bar{F}_s = \frac{I_s}{t_s} = 2p_e Q_e \; \frac{\frac{p_1}{p_e}}{(\frac{p_1}{p_e})^{\frac{1}{\gamma}} - 1} \; \sqrt{h^2 + k^2} \; (\sqrt{h^2 + k^2} - k).$$

The quantities h and k are defined in (7.2) and (7.3).

      Figures 7.6a,b give two sets of graphs of $\bar{F}_s/p_e Q_e$ as a function of p_1/p_e for various values of k (i.e. for various initial velocities U_1). For one set of graphs $\gamma = 1.4$, and for the other $\gamma = 1.2$. A comparison of (7.5) with (7.3) shows that $\bar{F}_s$ depends upon the fuel parameter $\frac{M_g}{M_w}$ and the initial velocity U_1 only in the combination given in the formula (7.3) for the quantity k. Figure 7.7 therefore includes the effects of both fuel parameter and initial velocity on $\bar{F}_s$ by giving $\bar{F}_s/p_e Q_e$ as a function of k. Two sets of graphs are given: one each for $\gamma = 1.4$ and $\gamma = 1.2$ and for the values $p_1/p_e = 10, 40, 100$.

      As we would expect, Figures 7.6a, b show that the average thrust $\bar{F}_s$ increases steadily with increase of the pressure ratio. The increase is rather rapid at first. But for $p_1/p_e > 40$, say, $\bar{F}_s$ increases nearly linearly with p_1/p_e.

      From Figures 7.7a and 7.7b we can deduce the effect of changes in either the fuel parameter $\frac{M_g}{M_w}$ or the initial velocity U_1 on $\bar{F}_s$, since these two quantities occur in (7.5) only through the quantity k given by (7.3). As $\frac{M_g}{M_w} \to 0$, $k \to \infty$, while

CONFIDENTIAL

$k \to 0$ when $\frac{M_g}{M_w} \to \infty$. The quantity k is linear in U_1. As we see from Figures 7.7a and 7.7b, $\bar{F}_s$ steadily decreases from a finite value at k = 0 and approaches asymptotically another constant non-zero value (which can be readily proved from 7.5 to be just half the value at k = 0) as k tends to ∞. In other words, the values of $\bar{F}_s$ lie between two finite bounds (with the lower bound half the upper bound) for all values of the fuel parameter M_g/M_w and the initial velocity U_1. This is a rather remarkable result. It says, for example, that the quantity of fuel injected per stroke can be decreased indefinitely and the average thrust over the stroke will not tend to zero, but will tend rather to a limit which has half the value of the maximum which is approached when the quantity of fuel injected is made large without limit.[*] This result is made somewhat less surprising when we recall that both the impulse per stroke I_s and the time of stroke t_s tend to zero when M_g/M_w tends to zero: apparently they approach zero in such a way that the ratio I_s/t_s tends to a finite and non-zero limit in general. It should also be recalled that the pressure ratio is assumed always to be fixed as the limit is approached, and, as in a similar case above, it is far from clear that M_g/M_w can be made very small while p_i/p_e is maintained fixed. In other words, it seems likely that the pressure ratio which can be achieved on injection of fuel will tend to zero with the quantity of fuel injected, and so the average thrust would also tend to zero through failure to maintain a minimum pressure ratio. Nevertheless, in view of the fact that the efficiency as measured by the effective velocity I_s/M_g has its maximum for $M_g/M_w = 0$, the following conclusion is strongly indicated: The hydropulse should be operated at high frequency

---

[*] It is to be remembered that the average thrust in question here refers to the average over the power stroke alone and not over the full cycle.

CONFIDENTIAL

with injection of small quantities of fuel per stroke. The quantity of fuel injected should be kept as low as is compatible with maintenance of a minimum pressure ratio (of the order of 30 to 40 if possible.)

The actual numerical values taken on by $\bar{F}_s/p_e Q_e$, as indicated in Figures 7.7a,b are also not without interest. These values are all seen to lie between 1.75 and 12, depending on γ and the initial pressure ratio (which we have assumed to be between 10 and 100.) We recall (cf. the discussion in Section 3 on the meaning of the quantities I and F) that the quantity $\bar{F}_s$ furnishes the average thrust (over the power stroke only) that is available for overcoming the total drag on the hydropulse. If the drag is given, the graphs 7.7a,b make it possible to estimate the minimum cross-section area of the hydropulse that would be necessary to overcome it. For example, suppose that a drag of 1500 pounds is to be overcome by a hydropulse which is operating quite near the surface of the water so that p_e = 15 pounds per square inch. Suppose, in addition, that time t_s for the power stroke is only 1/3 of the full cycle (i.e. the scavenging and refill parts of the cycle require twice as much time as the power stroke.) If γ = 1.2 and the pressure ratio p_i/p_e is about 40, then $\bar{F}_s/p_e Q_e$ = 3 approx. (cf. Figure 7.7a). The average thrust $\bar{T}_s$ over the full cycle would therefore be given by $\bar{T}_s/p_e Q_e$ = 1. For the cross-section area Q_e of the hydropulse, 100 square inches would then be required for a thrust of 1500 pounds. In other words, a cylindrical tube more than 10 inches in diameter would be required.

d) **Gas pressure and exit velocity as functions of the time.**

Our approximate formulas for the impulse, time of stroke, and average thrust, as discussed above, were all derived on the basis of results obtained through integration of the

56

equation of motion of the water in the hydropulse. For this
reason, and also because some of the results have an interest
for their own sake, we discuss briefly the integration of the
equation of motion (3.6) for the case of the tube of constant
cross-section.

In terms of the dimensionless variables introduced in
Section 4, the differential equation to be integrated is

$$(7.6) \qquad v_s^\gamma v^{-\gamma} - 1 - (1-v) \frac{d}{dv} (v')^2 = 0 \, ,$$

since in this case the quantity $q = Q/Q_e$ is identically one.
The prime, we recall, refers to differentiation with respect
to our dimensionless time τ. Since $\frac{d}{dv}(v')^2 = 2v''$ we may
write (7.6) in the form

$$(7.7) \qquad 2(1-v) \, v'' = v_s^\gamma v^{-\gamma} - 1 \, .$$

This differential equation is to be integrated for the initial
conditions

$$(7.8) \qquad \begin{cases} v(0) = v_1 \\ v'(0) = v_1' \, , \end{cases}$$

i.e. for a given initial gas volume and its rate of expansion.
The latter quantity is fixed by the initial velocity of the
water in the tube at the moment fuel is injected; it corresponds, therefore, roughly to the forward velocity of the
hydropulse since the water in the tube at the end of the refill
portion of the cycle would have about this velocity. The
quantity vi is fixed by the initial pressure of the gas once
its temperature and molecular weight are known. Once v has
been determined from (7.7) and (7.8) as a function of τ, we

57

CONFIDENTIAL

know that all quantities of interest, such as the impulse,
velocity of the water, etc. can be computed at once.

Our problem contains four parameters: $v_s = V_s/V_e$,
that is, the ratio of the gas volume at pressure p_e to the
total volume of the tube, γ the adiabatic exponent, and
vi and v_i'.

The integrations have been carried out numerically in
quite a large number of cases by the methods discussed in
Appendix II of this report. We report here on a few typical
cases in the form of graphs. In the cases we report the
adiabatic exponent has the value $\gamma = 1.3$. The quantity v_i
has the value .0256 -- that is, the initial volume of the gas
is chosen somewhat less than 3 per cent of the total volume
of the tube. The initial pressure ratio of the gas p_i/p_e is
40 and this, together with the adiabatic gas law in the form

$$(7.9) \qquad pv^{\gamma} = p_1 v_1^{\gamma} = p_e v_s^{\gamma} ,$$

has the effect of fixing the quantity v_s in (7.7).

The results of the numerical integration are shown in
the graphs of Figures 7.8, 7.9, and 7.10. Figure 7.8 gives
the "excess pressure ratio" $p/p_e - 1$ as a function of the "time"
τ for $v_i = 0, 1, 3.$[*]

It might be noted that the thrust F at any time is
given in this case by the formula

$$(7.10) \qquad F = p_e Q_e (p/p_e - 1) ,$$

so that Figure 7.8 yields $F/p_e Q_e$ as a function of τ. The

_____

[*] If p_e is atmospheric pressure, these values of v_i correspond
to forward velocities of the hydropulse of 0, 31.8, and
95.4 miles per hour.

CONFIDENTIAL

important observation to be made from the graph is that the
pressure falls off with extreme rapidity initially[*] -- a fact
which has been repeatedly stated. We observe also that the
rate of fall increases with increase in v_1', that is, with
increase in the initial velocity of the water in the tube.
For higher pressure ratio also the rate of decline increases
(though we do not reproduce here our results for higher
pressure ratios.) Figure 7.8 indicates why our approximate
formulas based on the assumption of impulsive action of the
injected gas on the water are quite accurate. Figure 7.9
makes this even more plausible. In these graphs the rate of
expansion of the gas is shown as a function of the gas volume,
which is the same effectively as giving the velocity of the
water as a function of the gas volume. As we see, the velocity
goes up very quickly at the beginning of the stroke and then
remains nearly constant over the remainder of the stroke.
The rate of approach to a nearly constant velocity is faster
with the higher initial velocities. From these observations
on the behavior of the gas pressure and the velocity of the
water in the early part of the stroke we see that the accuracy
of our approximate formulas is better the higher the initial
pressure of the gas and the initial velocity of the water are.
        Figure 7.10 is a graph showing the gas volume as a
function of the time. As we see, the gas volume is nearly
linear in the time except near $\tau = 0$ where the curves all
have high curvatures. The method of numerical integration
of the equation of motion explained in Appendix II is based
on this observed behavior of v as a function of τ.

---

[*] In Sections 9 and 10 of this report two possible means
of smoothing out these initial sharp variations in pressure
are discussed.

e) <u>Effect of linear buildup (in time) of the gas pressure.</u>

In all but this part of the present section we assume that injection of the fuel leads instantaneously to the initial gas pressure. We shall investigate briefly here the motion of the water in the tube which results when the gas pressure is assumed to increase linearly in the time from the value p_e to the value p_1. In this way we attempt to simulate what might occur in practice upon injection of a chemical into the tube. We obtain graphs from which the values of $v(\tau_0)$ and $v'(\tau_0)$ can be obtained once the pressures p_1 and p_e and the "time" τ_0 required to attain the pressure p_1 are given. Clearly, when $v(\tau_0)$ and $v'(\tau_0)$ are once known our usual methods and results for the case of instantaneous creation of the initial pressure can be used.

We proceed to derive the appropriate differential equation, making use of the dimensionless variables used above (Section 7d). In view of (7.7) and (7.9) we have

(7.11) $$v''(1-v) = \frac{1}{2}(p/p_e - 1) ,$$

in which p/p_e is assumed to be a linear function of τ, as follows:

(7.12) $$p/p_e = 1 + 2c\tau , \quad 0 \leq \tau \leq \tau_0 .$$

Thus (7.11) can be written

(7.13) $$v''(1 - v) = c\tau .$$

As initial conditions we have $v(0) = 0$, naturally, and we take $v'(0) = 0$ also (as was indicated above.) If we were to develop the solution of (7.13) in a power series under these initial conditions one would find readily that the series begins with the term in τ^3:

$$(7.14) \qquad\qquad v = a_3 \tau^3 + \ldots$$

This being the case it is convenient to introduce a new independent variable y, replacing τ, by the relation

$$(7.15) \qquad\qquad y = \frac{1}{6} c \tau^3 .$$

In this variable the differential equation (7.13) is

$$(7.16) \qquad\qquad (3y \frac{d^2 v}{dy^2} + 2 \frac{dv}{dy}) (1 - v) = 2 ,$$

and the initial conditions become

$$(7.17) \qquad\qquad \left[\begin{array}{l} v = 0 , \text{ and} \\ \frac{dv}{dy} = 1 \end{array} \right.$$

at $y = 0$, the condition on $\frac{dv}{dy}$ at $y = 0$ resulting from (7.16) at $y = 0$, in view of (7.14). By solving this one initial value problem we are enabled to determine v and v' as functions of y and, since the constant c does not occur in (7.16) or (7.17), we also obtain at one stroke the solutions of (7.13) for any value of c.

Suppose that the pressure ratio p/p_e is to have the value p_1/p_e at a specific time τ_o. The relation (7.12) would then determine the constant c:

$$(7.18) \qquad\qquad c = \frac{\frac{p_1}{p_e} - 1}{2 \tau_o} .$$

Our solution of (7.16) would then yield through (7.15) values for v and v' at $\tau = \tau_o$ and from these, in turn, our usual methods would determine the motion over the remainder of the stroke.

The two graphs shown in Figure 7.11 were drawn to facilitate such calculations: c is determined from (7.18) and from it and τ_o the value of y_o is fixed by (7.15). From the graphs the values of $v(\tau_o)$ and $v'(\tau_o)$ can then be read off.[*]

It is of some interest to consider a few numerical cases. We assume $\gamma = 1.4$ and $\tau_o = 0.1$. Two different pressure ratios pi/pe are taken, i.e. $pi/pe = 20$ and $pi/p_e = 40$. That is, the gas pressure ratio builds up linearly to a value of 20 or of 40 in the "time" $\tau_o = 0.1$. The following table summarizes the results obtained for the entire stroke, which comprises both the time during which the pressure ratio builds up from unity to its maximum and the time during which the gas expands adiabatically down to the exit pressure p_e. Thus the total impulse j_s, the time of stroke τ_s, and the dimensionless energy ϵ_s (imparted to the water by the expanding gas) are made up of two contributions

$$j_s = j_o + j_1 ,$$

$$\tau_s = \tau_o + \tau_1 ,$$

$$\epsilon_s = \epsilon_o + \epsilon_1 ,$$

---

[*] It should be recalled that these curves refer only to the case in which the water in the tube is initially at rest. There would be no difficulty in treating the cases with $U_i \neq 0$, but the calculations would be somewhat lengthier than those for $U_i = 0$.

CONFIDENTIAL

distinguished by subscripts, the subscript zero[*] referring
to the period of linear increase in pressure. (The quantity
ϵ_s is defined by $\epsilon_s = \dfrac{E_s}{p_e V_e}$, where E_s is given in (5.1).)

Linear Pressure Increase

$\dfrac{p_1}{p_e}$	j_o	j_1	$j_s = j_o + j_1$	τ_o	τ_1	$\tau_s = \tau_o + \tau_1$
20	.475	.274	.749	0.1	.158	.258
40	.975	.648	1.623	0.1	.256	.356
$\dfrac{p_1}{p_e}$	ϵ_o	ϵ_1	$\epsilon_s = \epsilon_o + \epsilon_1$	$v(\tau_o)$	$v'(\tau_o)$	m_g
20	.226	.336	.562	.0158	.475	.317
40	.951	1.696	2.647	.0325	.975	1.300

The quantities $v(\tau_o)$ and $v'(\tau_o)$ are the dimensionless gas
volume and rate of expansion at the time $\tau = \tau_o$. The quantity
m_g is the product $m_g = vi \cdot p_1/p_e$ of the volume vi and the
maximum pressure ratio; these numbers are computed in order
to make comparative estimates of the quantities of fuel
necessary in these cases and some others to be discussed shortly.
Clearly the quantities m_g are proportional to the masses of
gas present if p_e and the temperature and molecular weight of

---

[*]
The quantity j_o is given by: $j_o = \dfrac{1}{2} \displaystyle\int_o^{\tau_o} (\dfrac{p}{p_e} - 1) d\tau = \dfrac{c}{2} \tau_o^2$.
We determine j_1 by our usual methods from the initial
conditions $vi = v(\tau_o), v_1' = v'(\tau_o)$.

the gases are the same in the cases to be compared (assuming, as always, ideal gases.) We observe that the time $\tau_0 = 0.1$ assumed by us turns out to be about 40 per cent of the entire stroke in one case ($p_1/p_e = 20$) and about 30 per cent in the other (when $p_1/p_e = 40$). However, the impulse delivered during the time of increasing pressure is a considerably larger fraction of the total impulse over the fuel stroke -- it is about 65 per cent when $p_1/p_e = 20$ and about 60 per cent when $p_1/pe = 40$. We conclude: If the time required for the gas to reach its maximum pressure is an appreciable fraction of the total time of stroke, this period is of importance because the impulse delivered then is likely to be a large fraction of the total impulse.

,We proceed to make a comparison with a few cases in which the initial gas pressure is assumed to reach its maximum instantaneously. On what basis the comparison should be made is not entirely clear. We have chosen to make a comparison in two different ways: 1) We assume that p_1/pe and $v_1 = v(\tau_0)$ have the values given in the above table, and compute j_s and τ_s by our usual methods for instantaneous rise of pressure, with the additional assumption that $v_1' = 0$ (that is, the water is initially at rest) at $\tau = 0$. If we assume that the temperature and molecular weight of the gases are the same at the instant when their pressures are maximum, it follows that the masses of gas are the same in the cases to be compared.

p_1/p_e	τ_s		j_s	
	Linear Increase	Instantaneous	Linear Increase	Instantaneous
20	.258	.410	.749	.575
40	.356	.318	1.623	1.281

p_1/p_e	ϵ_s		m_g	
	Linear Increase	Instantaneous	Linear Increase	Instantaneous
20	.562	.336	.317	.317
40	2.647	1.696	1.300	1.300

These calculations show that a linear buildup of pressure
may be favorable for the operation of the hydropulse, as
compared with the instantaneous creation of the peak gas
pressure. The impulse delivered per stroke for a given
quantity of fuel is considerably greater, about 30 per cent,
when the pressure increases gradually. At the same time we
note the marked differences in the time of stroke τ_s.

The above comparison is based on assuming equality of
temperatures at the instant when the gas pressures are at a
maximum. But when the gas pressure builds up gradually, a
considerable amount of work is done before the gas attains
its peak pressure, and consequently the assumption of equality
of gas temperatures may be unreasonable. 2) In the second
mode of comparison, we assume the same values for p_1/p_e and
for the total energy ϵ_s imparted to the water. The results
are tabulated below:

p_1/p_e	τ_s		j_s	
	Linear Increase	Instantaneous	Linear Increase	Instantaneous
20	.258	.261	.749	.740
40	.356	.393	1.623	1.585

p_1/p_e	ϵ_s		v_1	
	Linear Increase	Instantaneous	Linear Increase	Instantaneous
20	.562	.562	.0158	.0264
40	2.647	2.647	.0325	.0507

It appears that the gradual buildup of pressure produces almost
the same total impulse as the instantaneous case. The
approximate formula

$$I_s = \sqrt{2M_w E_s}$$

derived in Section 5 is thus found to be accurate even in
the event of a gradual buildup of pressure. (The reason
for this is that the mass of water in the tube remains
sensibly constant during the buildup.)

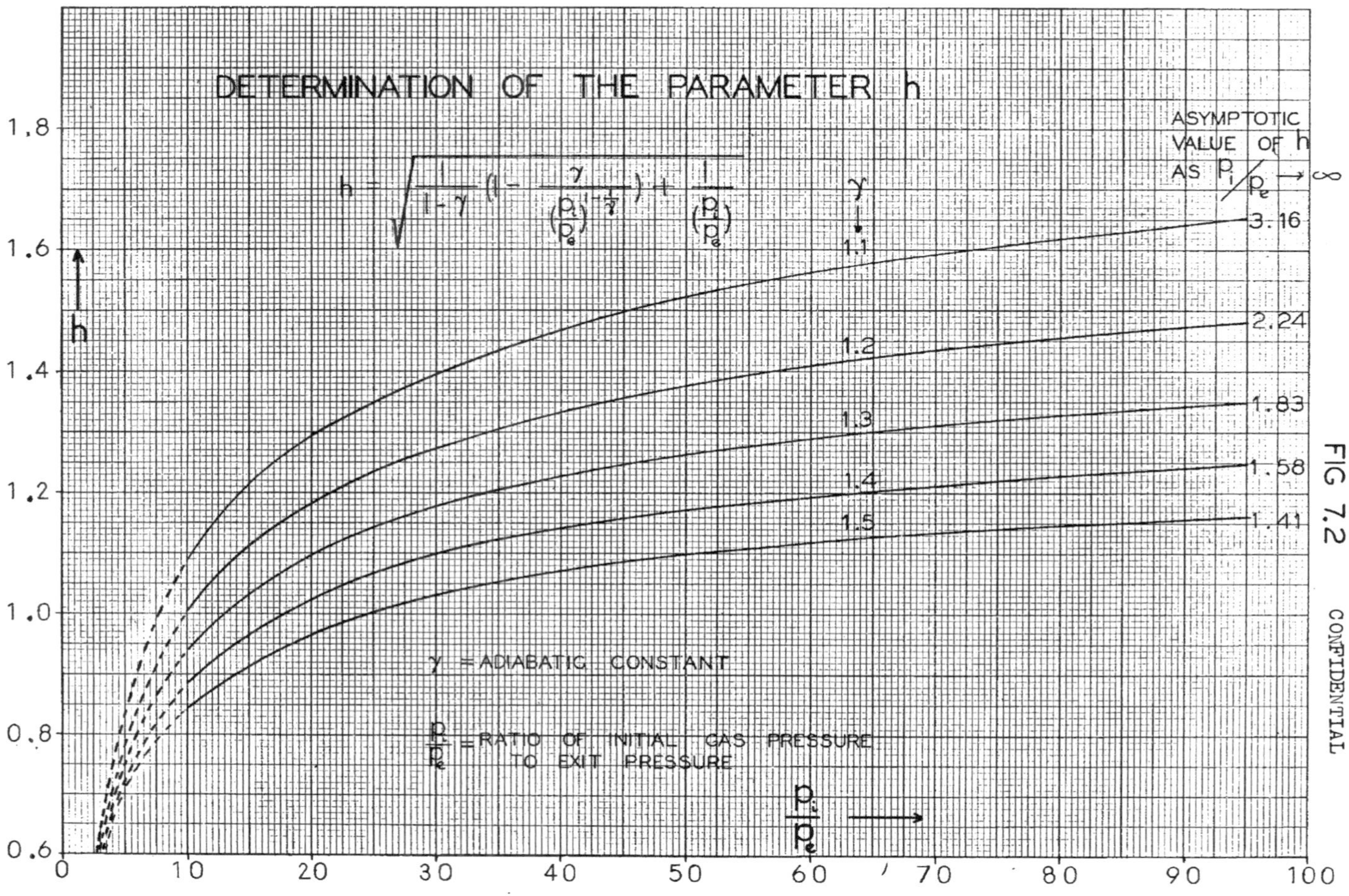
DETERMINATION OF THE PARAMETER h
$h = \sqrt{\frac{1}{1-\gamma}\left(1 - \frac{\gamma}{\left(\frac{p_i}{p_e}\right)^{1-\frac{1}{\gamma}}}\right) + \frac{1}{\left(\frac{p_i}{p_e}\right)}}$
ASYMPTOTIC VALUE OF h AS $p_i/p_e \rightarrow \infty$
γ
1.1
1.2
1.3
1.4
1.5
3.16
2.24
1.83
1.58
1.41
γ = ADIABATIC CONSTANT
$\frac{p_i}{p_e}$ = RATIO OF INITIAL GAS PRESSURE TO EXIT PRESSURE
h
$\frac{p_i}{p_e}$
1.8
1.6
1.4
1.2
1.0
0.8
0.6
0
10
20
30
40
50
60
70
80
90
100

FIG 7.2

CONFIDENTIAL

FIG 7.3   CONFIDENTIAL   71

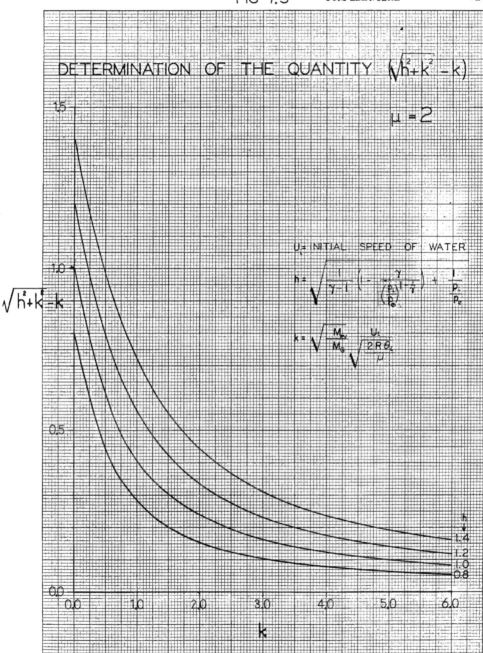

DETERMINATION OF THE QUANTITY $(\sqrt{h^2+k^2}-k)$

$\mu = 2$

$U_c = $ INITIAL SPEED OF WATER

$$h = \sqrt{\frac{1}{\gamma-1}\left(1 - \left(\frac{p_1}{p_2}\right)^{1-\frac{\gamma}{\gamma}}\right) + \frac{1}{\frac{p_1}{p_2}}}$$

$$k = \sqrt{\frac{M_w}{M_G}} \sqrt{\frac{U_c}{2R\theta_c}}$$

$\sqrt{h^2+k^2}-k$

h
1.4
1.2
1.0
0.8

k

FIG 7.4   CONFIDENTIAL   73

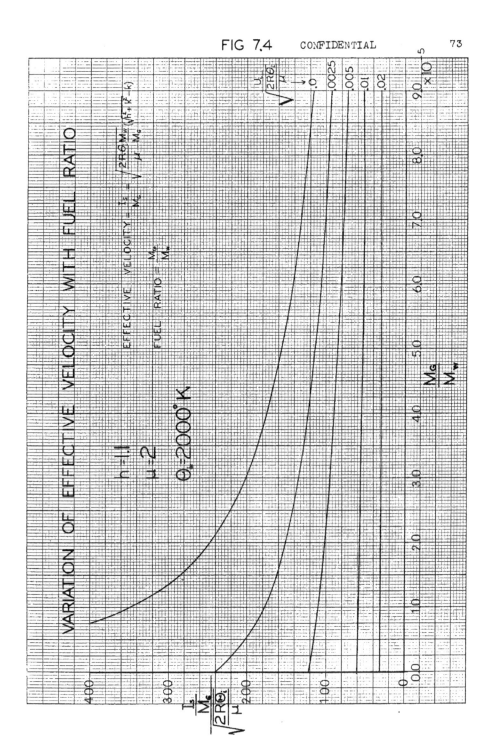

VARIATION OF EFFECTIVE VELOCITY WITH FUEL RATIO

$$\text{EFFECTIVE VELOCITY} = \frac{I_s}{M_c} = \sqrt{\frac{2R\theta_\circ}{\mu}} \frac{M_w}{M_G} \left(\sqrt{h^2 + k^2} - k\right)$$

$$\text{FUEL RATIO} = \frac{M_G}{M_w}$$

$$h = 1.1$$
$$\mu = 2$$
$$\theta_\circ = 2000°K$$

FIG 7.5a   CONFIDENTIAL   75

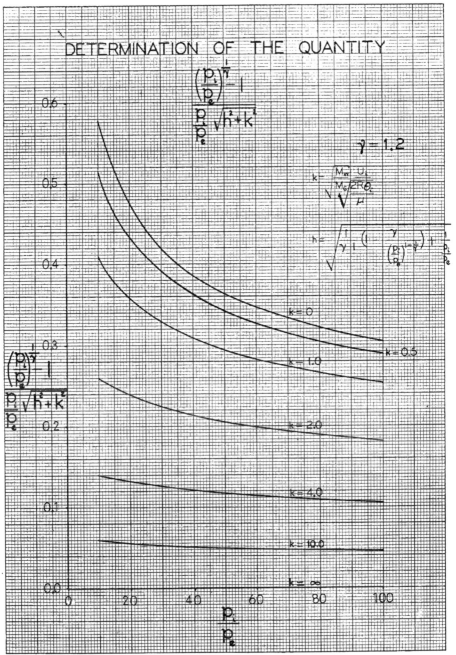

DETERMINATION OF THE QUANTITY

$$\frac{\left(\frac{p_i}{p_e}\right)^{\frac{1}{\gamma}} - 1}{\frac{p_i}{p_e}\sqrt{h^2 + k^2}}$$

$\gamma = 1.2$

$$k = \frac{M_{\overline{w}}}{M_G} \sqrt{\frac{U_i}{\frac{2R\theta_i}{\mu}}}$$

$$h = \sqrt{\frac{1}{\gamma - 1}\left(1 - \frac{\gamma}{\left(\frac{p_i}{p_e}\right)^{1-\frac{1}{\gamma}}}\right) + 1 - \frac{1}{\frac{p_i}{p_e}}}$$

$k = 0$

$k = 0.5$

$k = 1.0$

$k = 2.0$

$k = 4.0$

$k = 10.0$

$k = \infty$

FIG 7.5b   CONFIDENTIAL   77

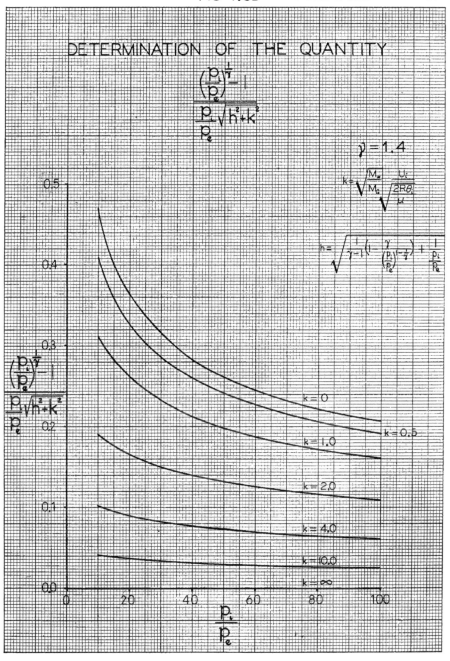

FIG 7.6a   CONFIDENTIAL   79

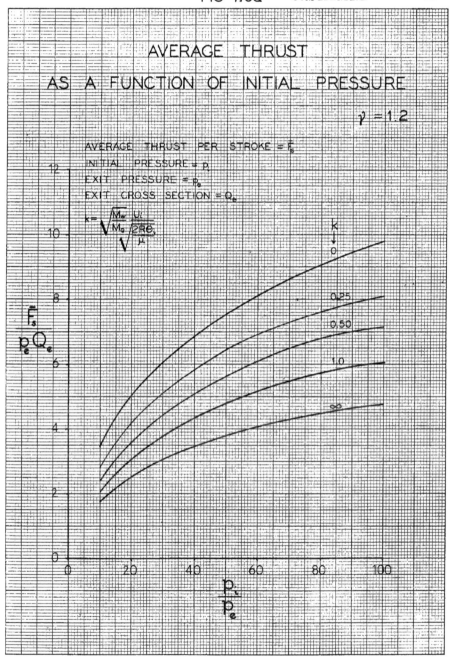

AVERAGE THRUST

AS A FUNCTION OF INITIAL PRESSURE

$\gamma = 1.2$

AVERAGE THRUST PER STROKE $= \bar{F}_s$
INITIAL PRESSURE $= p_i$
EXIT PRESSURE $= p_e$
EXIT CROSS SECTION $= Q_e$

$$k = \sqrt{\frac{M_w}{M_a}} \frac{U_i}{\sqrt{\frac{2R\theta_i}{\mu}}}$$

FIG 7.6b   CONFIDENTIAL   81

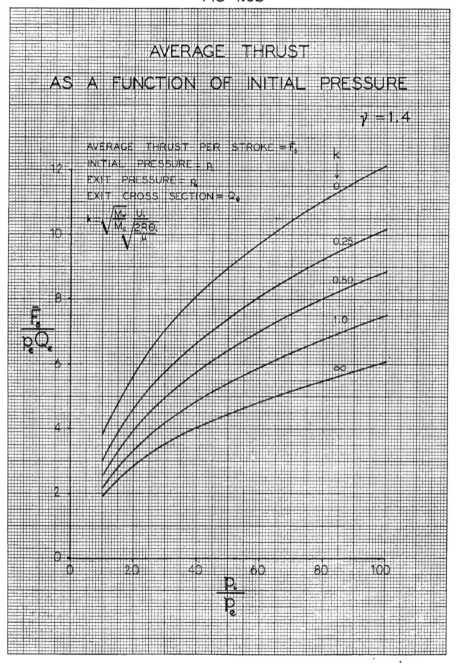

AVERAGE THRUST

AS A FUNCTION OF INITIAL PRESSURE

$\gamma = 1.4$

AVERAGE THRUST PER STROKE = $\bar{F}_s$
INITIAL PRESSURE = p_i
EXIT PRESSURE = p_e
EXIT CROSS SECTION = Q_e

$$k = \sqrt{\frac{M_e}{M_G}} \cdot \frac{U_L}{\sqrt{\frac{2R\Theta}{\mu}}}$$

k
0
0.25
0.50
1.0
∞

FIG. 7.7a   CONFIDENTIAL   83

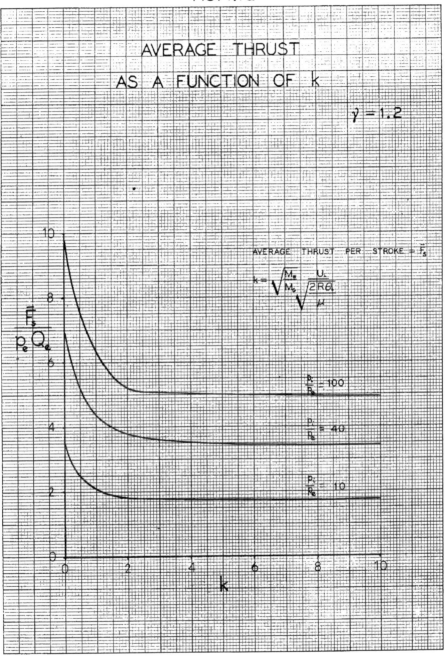

AVERAGE THRUST

AS A FUNCTION OF k

$\gamma = 1.2$

AVERAGE THRUST PER STROKE = $\bar{F}_s$

$$k = \sqrt{\frac{M_w}{M_c}}\sqrt{\frac{U_c}{2R\theta}}{\mu}$$

$\dfrac{\bar{F}_s}{p_e Q_e}$

$\dfrac{p_c}{p_e} = 100$

$\dfrac{p_c}{p_e} = 40$

$\dfrac{p_c}{p_e} = 10$

k

FIG 7.7b CONFIDENTIAL 85

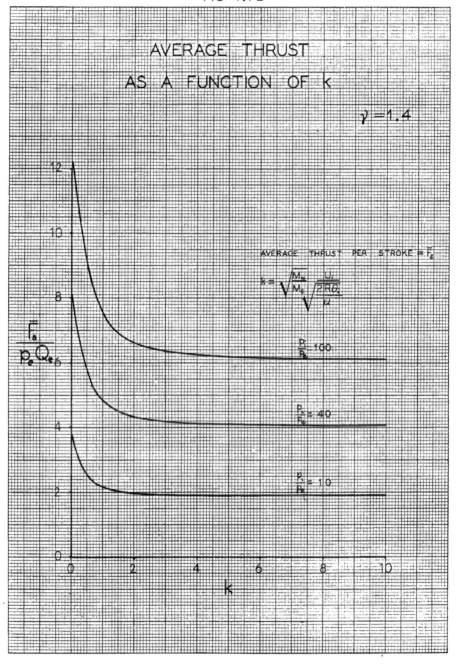

AVERAGE THRUST

AS A FUNCTION OF k

$\gamma = 1.4$

AVERAGE THRUST PER STROKE $= \bar{F}_s$

$$k = \sqrt{\frac{M_N}{M_0}} \sqrt{\frac{U_i}{\frac{2R\theta_i}{\mu}}}$$

$\frac{p_i}{p_0} = 100$

$\frac{p_i}{p_0} = 40$

$\frac{p_i}{p_0} = 10$

$\frac{\bar{F}_s}{p_0 Q_0}$

k

FIG 7.8   CONFIDENTIAL   87

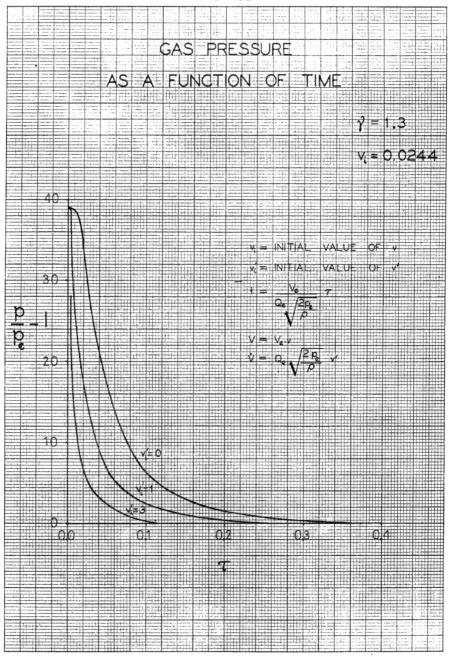

GAS PRESSURE

AS A FUNCTION OF TIME

$\bar{\gamma} = 1.3$

$v_i = 0.0244$

$v_i = $ INITIAL VALUE OF v

$v_i' = $ INITIAL VALUE OF v'

$t = \dfrac{V_e}{Q_e \sqrt{\dfrac{2 p_e}{\rho}}} \, \tau$

$V = V_e v$

$\dot{V} = Q_e \sqrt{\dfrac{2 p_e}{\rho}} \, v'$

$\dfrac{p}{p_e} - 1$

$v' = 0$

$v' = 1$

$v' = 3$

τ

FIG 7.9   CONFIDENTIAL   89

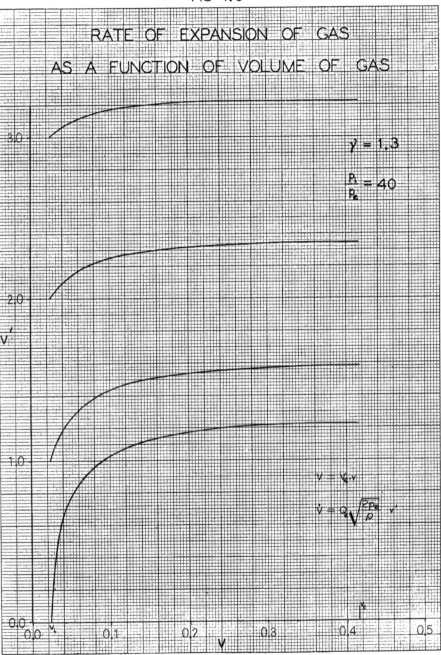

RATE OF EXPANSION OF GAS
AS A FUNCTION OF VOLUME OF GAS

$\gamma = 1.3$

$\dfrac{P_1}{P_2} = 40$

$v = V_b \cdot v'$

$v = a_2 \sqrt{\dfrac{\rho p_2}{\rho}} \ v'$

FIG 7.10   CONFIDENTIAL   91

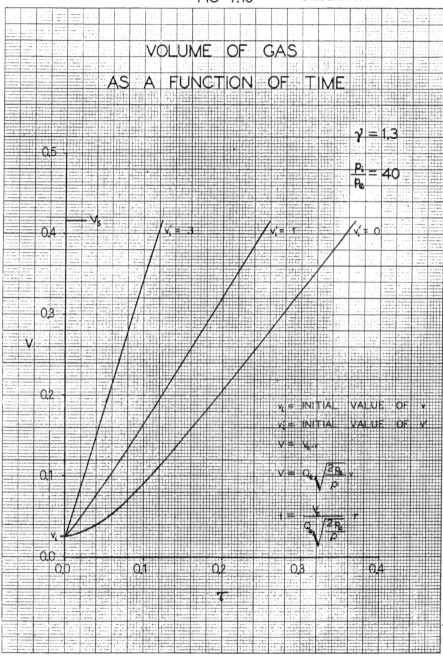

FIG 7.11   CONFIDENTIAL   93

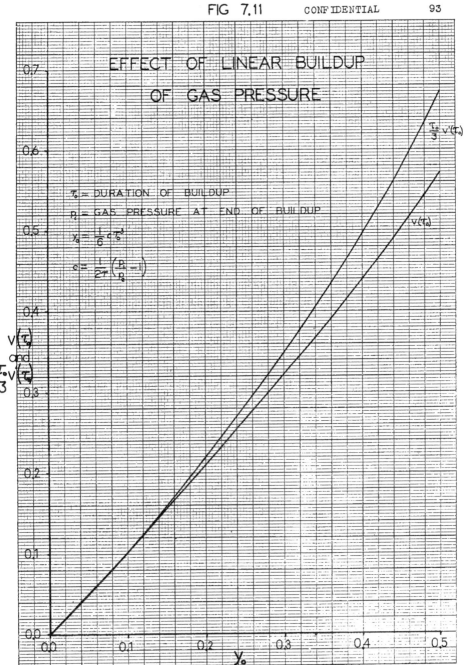

EFFECT OF LINEAR BUILDUP
OF GAS PRESSURE

τ_0 = DURATION OF BUILDUP

p_1 = GAS PRESSURE AT END OF BUILDUP

$x_0 = \frac{1}{6} c \tau_0^3$

$c = \frac{1}{2\tau} \left(\frac{p_1}{p_0} - 1 \right)$

Section 8. Tubes with conical portions.

We have seen in Section 5 that the impulse per stroke
is not sensitive to changes in shape of the hydropulse tube
if the initial velocity of the water in the tube is not great.
However, if the initial velocity of the water is high, it is
possible that the impulse may be considerably affected by
changes in shape of the tube. In this section we give the
results of calculations for the impulse and time of stroke for
three different shapes of tubes with front portions made up
of conical pieces joined to a cylindrical rear portion. These
three cases are compared with that of the cylindrical tube of
uniform section. The calculations in some of the cases were
carried out in two different ways -- once by numerical integration of the differential equation of motion and once through
use of the approximate formulas developed in Section 5, so
that some idea of the accuracy of these formulas for tubes
of non-uniform section is obtained.

We make use of the dimensionless variables defined in
Section 4. The four tubes we consider are then defined by
giving the dimensionless cross-section area q as a function
of the dimensionless gas volume v, as follows (the accompanying
sketches show the tubes in the correct proportions):

<u>Case a)</u>

a)
$$q = \begin{cases} (v + \frac{1}{2})^{2/3} \,, & 0 \leq v \leq \frac{1}{2} \\ 1 \,, & \frac{1}{2} \leq v \leq 1 \end{cases}$$

CONFIDENTIAL

Case b)

b)
$$q = \begin{cases} (-2v + 2)^{2/3} , & 0 \leq v \leq \frac{1}{2} \\ 1 , & \frac{1}{2} \leq v \leq 1 \end{cases}$$

Case c)

c)
$$q = \begin{cases} (2v + 1)^{2/3} , & 0 \leq v \leq \frac{1}{4} \\ (-2v + 2)^{2/3}, & \frac{1}{4} \leq v \leq \frac{1}{2} \\ 1 , & \frac{1}{2} \leq v \leq 1 \end{cases}$$

Case d) (The cylinder)

d)
$$q \equiv 1 .$$

CONFIDENTIAL

We have calculated the values of j_s and τ_s in each of these cases when $\gamma = 1.1$, $p_1/p_e = 74$, $v_i = .01$. Two different values for the initial rate of expansion of the gas v_1' were taken: $v_i' = 0$ and $v_i' = 3$. The first value corresponds to zero initial velocity of the water in all four cases. The second value corresponds in the Case d) of the straight cylindrical tube to a velocity of 95.4 mi./hr. if the exit pressure p_e is atmospheric pressure. In the other cases, the velocity of the water is different in different parts of the tube. Also, in Cases a) and b) the velocity at the entrance initially is not the same as it is in Cases c) and d), but is greater in Case a) and less in Case b) than in Cases c) and d) since the cross-section area at the entrance is less in Case a) and greater in Case b) than in the other two cases. Probably a more realistic comparison for Cases a) and b) would result if the initial expansion rate v_i' were adjusted so that the entrance velocities would be the same initially in all cases to be compared.

The following table summarizes the results of our calculations. The columns headed by j_s(approx.) and τ_s (approx.) refer to values obtained by use of the approximate formulas[*] of Section 5:

---

[*] The approximate value of τ_s was obtained using formula (5.19) with the equality sign (converted to dimensionless form.)

	Case	j_s	j_s(approx.)	τ_s	τ_s(approx.)
$v_1' = 0$	a)	1.361	1.373	.407	.373
	b)	1.367	1.373	.377	.360
	c)				
	d)	1.366	1.373	.392	.353
$v_1' = 3$	a)	.3013	.3053	.139	.138
	b)	.3587	.3595	.159	.158
	c)	.3303	.331	.154	.154
	d)	.3014	.3019	.150	.148

Upon comparing the values of j_s for the four different shapes of tube when $v_1' = 0$ (i.e. when the initial velocity of the water is zero) we see that they are practically the same in all cases. This is in accord with the remarks on this point in Section 5, where it was stated that the impulse per stroke is insensitive to changes in shape of the tube at low initial velocities of the water. However, when $v_1' = 3$, some considerable divergence in the values of j_s is noted: The impulse is 20 per cent greater for the tube of Case b) (with the flared out front section) than for the Case d) of the straight cylinder. However, as we have already remarked, the Cases b) and d) for $v_1' = 3$ may not be strictly comparable because the entrance velocities of the water would be considerably different initially. This criticism does not apply

to the comparison of Cases c) and d) when $v_1' = 3$; we note
that j_s is 10 per cent greater for Case c) than for the
straight tube of Case d). These calculations indicate therefore that the impulse per stroke may be appreciably affected
by the shape of the tube when the initial velocity of the
water (which is roughly the same as the forward velocity of
the hydropulse) is high. It is indicated that a flared-out
front section may have advantages from the point of view of
maximizing the impulse.

We see that the values of both j_s and τ_s are given
quite accurately by the approximate formulas when $v_1' = 3$.
The accuracy of these formulas is not as high for $vi = 0$,
especially for τ_s, but the error is still not unreasonably
high. We observe also that the approximate formulas yield
values for j_s which are too high and values for τ_s which
are too low -- thus verifying in these special cases the
general rule, given in Section 5, that these formulas yield
upper and lower bounds for j_s and τ_s respectively.

It is clear that the results of this section are
concerned with too restricted a range of parameter values to
be conclusive. More work should be done in this direction.
It should also be pointed out that the assumptions leading to
the hydraulic theory used throughout this report are probably
too crude to deal accurately with tubes of rapidly varying
cross-section area -- the flow problems become more acute in
these cases.

CONFIDENTIAL

Section 9. Straight tube with mass and spring for elastic energy storage.

As we have seen in Section 7, the gas pressure decreases very rapidly from its initial high value at the beginning of the stroke while the exit velocity rises very rapidly. It seems likely that efficiency would be promoted by smoothing out these initial abrupt variations in pressure and exit velocity. In this section we investigate the use of a mass and spring as one possible method of achieving a smoother operation of the hydropulse. That a mass and a mechanical spring would be used in actual practice is perhaps doubtful -- our results simply give an indication of the possibilities inherent in provisions for elastic storage of energy during the early part of the stroke. In the section following this one, another less flexible but perhaps more practical method of achieving the same kind of results is considered.

We consider the case of a cylindrical tube which is fitted at the end by a piston of mass M backed up by a spring, as indicated in the following figure:

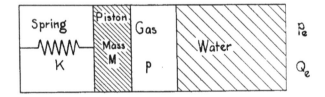

The fundamental assumptions regarding the behavior of the water
and gas to the right of the piston are the same as those made
in Section 2. Consequently the equation of motion of the water
in the tube is:

$$(9.1) \qquad -\frac{1}{Q_e^2}\, W\overset{\shortmid\shortmid}{W} = \frac{1}{\rho}(p - p_e),$$

in which W is the volume of water in the tube at any instant.
Equation (9.1) results from the fundamental differential
equation (3.6) of Section 3 when it is assumed that the tube
has a constant cross-section area Q_e throughout. The equation
of motion of the piston is given by

$$(9.2) \qquad -\frac{M}{Q_e^2}\, (\overset{\shortmid\shortmid}{W} + \overset{\shortmid\shortmid}{V}) = -\frac{K}{Q_e^2}\, (V_e - W - V) - p\ ,$$

in which V is the volume of gas at any instant, V_e is the
combined volume of gas and water when the spring is in its
unstrained position, and K is the spring constant. In addition
to (9.1) and (9.2) we used the relation between the gas pressure
P and its volume V which, as always, is taken as the adiabatic
relation pV^{γ} = const.

We introduce the dimensionless variables w, v, and τ
as in Section 4 and, in addition, the quantities

$$(9.3) \qquad \psi = \frac{M}{\rho V_e}\ , \quad \kappa = \frac{Q_e^2 p_e}{K V_e}\ ,$$

in place of the mass M and the reciprocal of the spring constant
K. The quantity ψ is the ratio of the mass of the piston to
that of the mass of water required to fill the volume V_e.
Equations (9.1) and (9.2) become in the new variables:

$$(9.4) \qquad\qquad - 2ww'' = v_s^\gamma v^{-\gamma} - 1$$

$$(9.5) \qquad\qquad - 2\mu(w'' + v'') = \frac{1}{\kappa}(1 - w - v) - v_s^\gamma v^{-\gamma} \; ,$$

in which the primes now refer to differentiations with respect to the "dimensionless time" τ and v_s is the value of v when the pressure P of the gas is p_e. As initial conditions (at the "time" $\tau = 0$) we prescribe, of course, the initial values of v, w and their first derivatives, or, what is the same thing, the initial position and velocity of the piston and the gas-water interface. Our initial conditions are, then:

$$(9.6) \qquad \begin{aligned} v\,(0) &= v_1 \; , \; w\,(0) = w_1 \; , \\ v'(0) &= v_1' \; , \; w'(0) = w_1' \; , \end{aligned}$$

in which v_1, v_1', w_1, and w_1', are given constants.

Once the solutions $v(\tau)$ and $w(\tau)$ of (9.4) and (9.5) satisfying the initial conditions (9.6) have been obtained, the thrust, gas pressure, impulse, exit velocity of the water, etc. can be calculated. Of particular interest are the thrust F and impulse per stroke I_s, which are given (in dimensionless form) by

$$(9.7) \qquad f = \frac{F}{2p_e Q_e} = \frac{1}{2\kappa}(1 - w - v) \; ,$$

$$(9.8) \qquad j_s = I_s / \rho V_e \sqrt{\frac{2p_e}{\rho}} = -\frac{1}{2} \int_0^{\tau_s} [\,\frac{1}{\kappa}(1 - w - v) + 1]\, d\tau \; .$$

The first formula results at once from the fact that the thrust in this case can be computed from the strain in the spring. The upper limit τ_s on the integral in (9.8) refers to the total time of the stroke, which is taken here, as in the preceding cases, as the time required for the gas to expand down to the exit pressure p_e.

The integration of the differential equations has been carried out for certain special initial conditions and special values for the parameters.[*] In all cases we have taken $\gamma = 1.5$ (which would be somewhat too high in practice, but a convenient and reasonable value for the purposes we have in mind) and $v_s = 0.5$, which means that the gas volume at pressure p_e is just one-half the volume of the tube when the spring is in its unstrained position. Except for one case to be discussed a little later, we take always as initial conditions $v_1' = w_1' = 0$, which means that the piston and the water in the tube are at rest initially, and $v_1 = 0.05$, $w_1 = 0.95$, which means that the initial volume of gas is 5 per cent of the volume of the tube. With $\gamma = 1.5$ one readily calculates the ratio p_1/p_e of the initial gas pressure to the exit pressure from the adiabatic relation and the values of v_s and v_1; the result is

(9.9) $$p_1/p_e = 31.6 .$$

For μ we have taken the value[**] 0.02, and for ψ the values 0.1, 0.4, 0.5, and 1.0.

---

[*] The differential equations appear not to be integrable by explicit means. The results given here were obtained by step-wise integrations, using the method of finite differences. The method of calculation is explained in Appendix I.

[**] This means that the exit pressure p_e acting on the piston would cause it to compress the spring an amount equal to 2 per cent of the length of the tube, as one can readily see (cf. (9.3)).

Figure (9.1) illustrates in a general way what occurs
when the gas is "injected" and the water and piston are set
in motion. The values of v, w, and p/p_e as functions of τ
are shown for the case $\psi = 0.1$ (mass of piston: mass of
water filling the whole tube = 0.1). One observes that w,
representing the volume of water in the tube, steadily decreases,
as it should since the water is always subjected to a resultant
force in the direction of the exit. The curve showing the
variation in v, which represents the gas volume, is not of so
simple a character. One observes that the gas volume first
increases during about one-third of the stroke from the initial
value 0.05 to a maximum, then decreases during the second
third of the stroke to a minimum, after which it rises to a
value slightly greater than 0.5 at the end of the stroke.
This means that the piston, which was at first set in motion
to the left with resulting compression of the spring, was
finally decelerated and brought to rest, after which it reversed
its motion and gathered enough speed to cause a compression
of the gas in spite of the fact that the gas-water interface
was moving always toward the right. This behavior of the
system is also brought out by the curve showing the pressure
ratio p/p_e as a function of the time: At first the gas pressure
drops very rapidly, then rises again when the gas is compressed
by the action of the piston, and finally falls to the exit
value.

Figures (9.2) and (9.3) give the pressure ratio and
the thrust f (cf. (9.7)) as functions of τ for various values
of ψ. For the larger values $\psi = 0.5$ and $\psi = 1.0$, i.e.
for the larger piston masses, the pressure decreases steadily,
but for the two smaller values it oscillates. This means
simply that the lighter pistons would oscillate rapidly enough
to cause a compression of the gas during the time of the stroke.
It is of some interest to observe that the pressure ratio in

the case $\psi = 0.4$ just failed to attain the value unity near $\tau = 0.6$. This suggests that for some value of ψ near 0.4 there would be a discontinuous change in the value for the time of stroke τ_s, when it is regarded as a function of ψ. Figure (9.3) shows the variation of the thrust over the stroke. We observe that the thrust actually becomes negative over part of the stroke in the cases $\psi = 0.1$ and $\psi = 0.4$. This means, of course, that the "thrust" on the hydropulse during this part of the stroke would actually be a drag.

Thus we see that the time variation in gas pressure and thrust can be changed quite radically by varying the mass of the piston, once a suitable spring has been chosen.[*] In particular, we observe from Figure (9.3) that the time of stroke seems to increase with decrease in mass of the piston, though this effect is not very pronounced.

The graphs shown in Figure (9.4) gives the thrust as a function of the time in a case which differs considerably from those just under discussion. In this case we suppose that the spring is compressed initially, and that the piston is moving toward the right, i.e. the compression in the spring is being relaxed, at the instant when the gas is injected. If the initial compression in the spring, the velocity of the piston, and the initial pressure of the injected gas are properly chosen one might expect a smoother distribution of the thrust over the stroke, since the thrust would begin with a positive value (instead of zero, as in the preceding cases) and could be maintained positive throughout the stroke. The graphs of Figure (9.4) bear this out rather strikingly. These graphs

---

[*] The value $\kappa = 0.02$ which characterizes the spring in our cases, was chosen so that the effects we seek would occur: It is clear that too stiff a spring would furnish essentially the same results as a rigid wall, while too weak a spring could not supply a thrust unless very large displacements could be permitted.

resulted from the following choice of parameter values:

$\psi = 1.0$ (a rather heavy piston, in other words), $\kappa = 0.02$, $v_s = 0.5$, vi $= 0.05$, $w_1 = 1.143$ ($v_1 + w_1 \neq 1$, as before, since the spring is initially compressed), $v_i' = 0$, $w_i' = -0.3375$ and -0.75 (that is, two different initial velocities were taken.) One observes that the case in which $w_1' = -0.3375$ yields a particularly smooth thrust curve.

Summing up, we might draw the following general conclusions:

1. Provisions for elastic energy storage can result in quite considerable changes in the distribution of the gas pressure and thrust during the time of the stroke.

2. The time of stroke τ_s may be a rather unstable quantity. As we have seen (in the case $\psi = 0.4$ above), the time of stroke may change discontinuously at certain parameter values. If the hydropulse were to be provided with a sensitive valve mechanism which would function only when the gas pressure is very near to p_e, it could happen that the actual operation of the hydropulse would be erratic if it were to take place near such unstable values of the parameters.

3. If gas injection, valve operation, and the functioning of elastic energy storage mechanisms can be properly timed, it appears possible to achieve quite smooth distribution of the thrust (and other quantities as well) over the stroke.

FIG. 9.1   CONFIDENTIAL   107

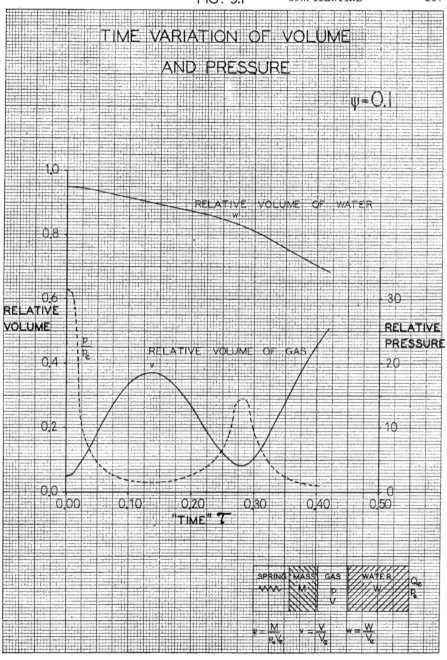

FIG. 9.2 CONFIDENTIAL 109

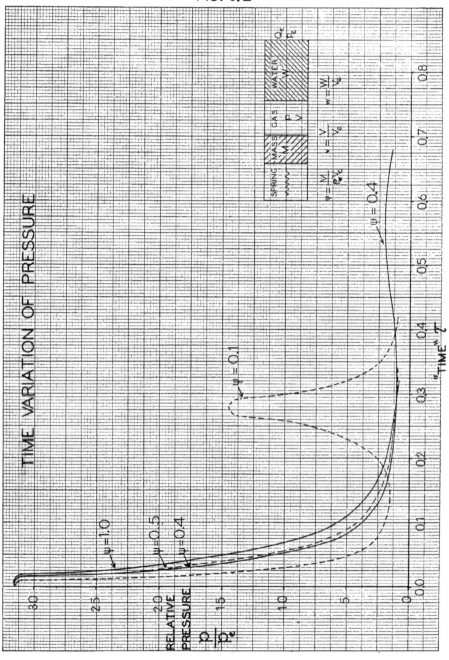

TIME VARIATION OF PRESSURE

FIG. 9.3   CONFIDENTIAL   111

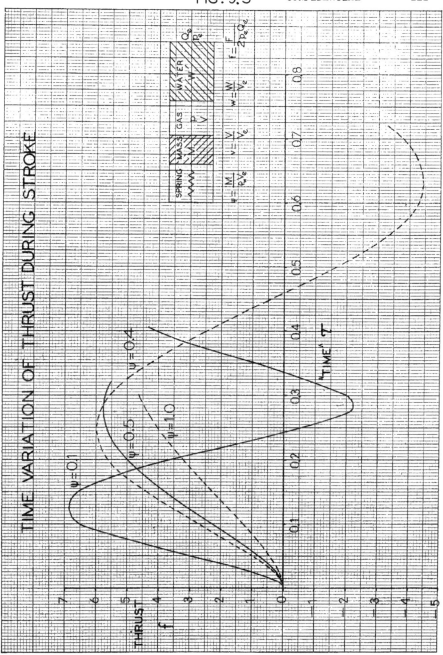

TIME VARIATION OF THRUST DURING STROKE

FIG. 9.4   CONFIDENTIAL   113

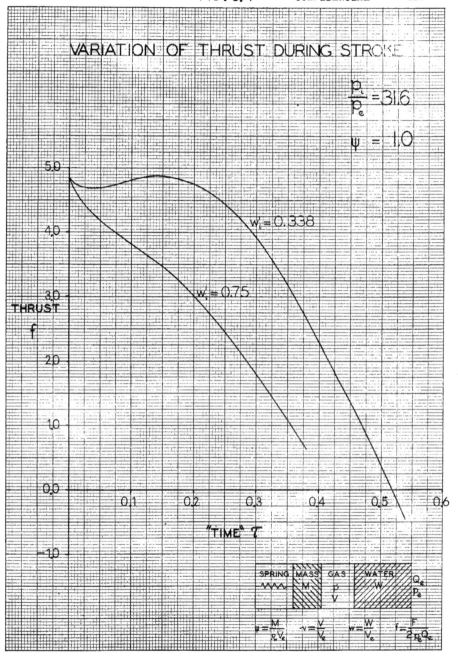

VARIATION OF THRUST DURING STROKE

$\frac{p_i}{p_e} = 31.6$

$\psi = 1.0$

$w_i = 0.338$

$w_i = 0.75$

THRUST
f

"TIME" τ

SPRING	MASS	GAS	WATER	
⋁⋀⋁⋀	M	$\frac{p}{V}$	W	$\frac{Q_e}{P_e}$

$\psi = \frac{M}{\rho V_e}$ $v = \frac{V}{V_e}$ $w = \frac{W}{V_e}$ $f = \frac{F}{2 p_e Q_e}$

Section 10. Spring effect obtained from a layer of gas.

In the preceding section we have seen that advantageous effects on the operation of the hydropulse might be achieved through the cushioning effect produced by use of a mass and a mechanical spring. It is not inconceivable that a mass and spring might be used in practice, but it would seem more reasonable to try to achieve similar effects through use of a layer (or layers) of gas as a "spring" together with a layer of water as mass. In this section we give the results of calculations for the case of a cylindrical tube into which two layers of gas are "injected" -- one at the closed end of the tube, the other further along the tube toward the exit section -- with a layer of water between the two layers of gas.[*] The situation is as indicated in the figure:

----

[*] In practice it is possible, even probable, that many such layers might co-exist in the tube at any instant. However, it is probably not worthwhile to undertake the very laborious calculations which would be required for such cases until more specific information on the possible ranges in the values of the parameters is available.

We make the same fundamental assumptions as in Section 2 regarding the behavior of the gas and water layers, but since we have to deal here with two different layers of both gas and water, it is necessary to supplement the quantities defined in Section 2 with a few others. We introduce the following volume ratios (cf. the above figure) as the principle dependent variables:

(10.1)
$$\xi = \frac{V_1}{V_e} \, ,$$

$$\eta = \frac{V_2}{V_e} \, ,$$

in which V_e is, as usual, the total volume of the tube. In addition, we introduce the quantity $\alpha = W_1/V_e$ for the relative volume of the layer of water between the two layers of gas. The equations of motion in terms of our dimensionless variables are than obtained for each of the layers of water in essentially the same manner as in Section 3. We obtain the following differential equations for the quantities ξ and η :

(10.2)
$$\xi_s^\gamma \xi^{-\gamma} - \eta_s^\gamma \eta^{-\gamma} - 2\alpha \xi'' = 0 \, ,$$

(10.3)
$$\eta_s^\gamma \eta^{-\gamma} - 1 - 2(\xi'' + \eta'')(1 - \xi - \eta - \alpha) = 0 \, ,$$

in which the primes mean differentiations with respect to the dimensionless time τ and the quantities ξ_s and η_s are the relative volumes of the gas layers when the pressure in these layers has the value p_e at the exit. We assume the gases to expand in accordance with the adiabatic relation $pv^\gamma = \text{const.}$, with $\gamma = 3/2$.

In addition to the differential equations, initial

conditions for the position and velocity of the masses in the
system must be imposed. In our numerical calculations we have
in all cases assumed the two masses of water to be initially
at rest; thus we have as initial conditions

$$(10.4) \qquad\qquad \xi'(0) = \eta'(0) = 0 ,$$

$$(10.5) \qquad\qquad \xi(0) = \xi_1; \eta(0) = \eta_1 ,$$

in which ξ_1 and η_1 are constants giving the initial values
of the relative volumes of the two gas layers.

Our system of equations contains five parameters:
ξ_1, η_1, ξ_s, η_s, and α. We do not propose to make
an exhaustive study involving variations of all of these parameters,
since our purpose here is simply to gain a qualitative impression
of the possibilities. Thus we limit ourselves in what follows
to the cases in which $\xi_s = \eta_s = \alpha = \frac{1}{4}$, that is, we assume
that the volumes of the gas layers at the pressure p_e are $\frac{1}{4}$
of the volume of the tube, and that the layer of water between
the two gas layers has the same volume. However, various
combinations of values of the initial relative volumes ξ_1
and η_1 have been taken, which amounts to saying that different
combinations of the initial pressure ratios p_i/p_e in each gas
layer were taken. The seven different combinations of values
used in the numerical calculations are indicated in the
following table:

$$\xi_1 = .025, .01 \quad \text{with no second layer}$$
$$\xi_1 = .025, \quad \eta_1 = .025, .01$$
$$\xi_1 = .01 , \quad \eta_1 = .025, .0156, .01.$$

The relations between these values for the volume ratios and

the corresponding values of the pressure ratios p/p_e is as
follows:

Volume ratios	Pressure ratios
.025	32
.0156	64
.01	125

The results obtained through numerical integration of
the differential equations[*] for all of these cases are given
in the form of graphs. Figures (10.1), (10.2), (10.3), and
(10.4) give the variation in pressure at the closed end of the
tube as a function of time for each of the seven different
combinations of initial pressure ratios. The pressure ratios
are plotted on a logarithmic scale, while the "time" τ is also
plotted logarithmically, on two of the sets of curves. The
principal observation to be made from these curves is that
considerable modifications in the shape of the early part of
the pressure-time curves can be achieved by taking various
combinations of initial pressure ratios in the two layers of
gas. Consider, for example, the combination of pressure ratios
32 - 125. In this case we observe that the pressure at the
closed end of the tube rises initially (instead of falling very
rapidly, as it does in most cases) because of the high pressure
in the second gas layer. However, the peak pressure ratio is
about 92, that is, considerably less than the initial high
value in the second gas layer. This shows clearly that the
smoothing out of the thrust curve in the early part of the
stroke is feasible through use of a cushioning gas layer. The
case 125 - 32 is also interesting: here the pressure curve
drops very rapidly at first, but reaches a minimum, goes up

---

[*] The method of finite differences was used. For a full
discussion see Appendix I of this report.

to a maximum and then falls steadily. This means that the second layer of gas was at too low a pressure initially to prevent the first layer from expanding rapidly. Instead, the second layer was compressed at first but later gave up some of its stored energy to compress the first layer slightly. One sees from these examples that the interactions between the two layers can be rather complicated.

Figure (10.5) is a graph of the (dimensionless) total impulse delivered to the tube up to any time τ, as a function of τ. That is, it is a graph of $j = \frac{1}{2} \int_{0}^{\tau} (p/p_e - 1)d\tau$ against τ, with p/p_e representing the pressure ratio of the gas layer at the closed end of the tube. One observes that the greater part of the impulse is delivered during the early part of the stroke. However, comparison of the case 32 - 0 with 32 - 32 and of the case 125 - 0 with 125 - 32 indicates that the introduction of the second layer has a slight smoothing effect.

It is found, as might have been expected, that the time of stroke τ_s (that is, the time required for the gas at the closed end to expand down to exit pressure) is increased by the introduction of a second gas layer, the increase being smaller as the pressure of the second layer goes up. The total impulse delivered over the stroke is also increased by the introduction of a second gas layer. However, the average thrust j/τ_s over the stroke does not always increase when a second gas layer is present. Our calculations indicate, in fact, that the average thrust may be decreased by the presence of a cushioning layer if its initial pressure is sufficiently low. These results are summarized in the following table:

Initial Pressure Ratios	τ_s	j	j/τ_s	$\dfrac{j}{m_{Fuel}}$ = Effective Velocity Ratio (based on 32-0 as unity)
125 - 0	0.205	1.34	6.54	0.91
125 - 32	0.257	1.59	6.19	0.66
125 - 64	0.273	1.69	6.19	0.64
125 - 125	0.254	1.77	6.97	0.60
32 - 0	0.290	0.93	3.21	1.00
32 - 32	0.370	1.29	3.48	0.70
32 - 125	0.342	1.57	4.59	0.65

The last column was obtained by assuming the gases in both layers to have the same initial temperature and molecular weight in all cases. These ratios are based on assuming the case 32 - 0 to be the standard, the other numbers representing the relative effective velocities (for both layers combined) in the other cases. The effective velocity increases as the mass of gas increases (as in Section 7.)

Figures (10.6) and (10.7) are the results of an attempt to compare the effects of a mass and a mechanical spring with those for a layer of gas used as a "spring". On what basis one should make the comparison is neither uniquely defined nor entirely clear. We have chosen to make the comparison assuming the masses of the piston and the water layer to be the same (that is, $\psi = 0.25$ for the spring case), while the spring constant was chosen to correspond to the "stiffness" of the gas layer at a pressure ratio of 31.6. In Figure (10.6) the pressure on the closed end of the tube is given as a function of τ for various initial conditions, with pressure in the case of the spring referring to spring thrust per unit cross-section area. The numbers (125 - 125) and (31.6 - 125) attached to the curves referring to the mechanical spring cases mean that the spring was initially compressed an amount which

would correspond to initial pressure ratios of 125 and 31.6
at the closed end. One observes that the curves for the
comparable cases coincide fairly closely in the initial stages
of the motion, but differ widely during the last 70 - 80 per
cent of the stroke. The curves for the mechanical spring cases
show rather heavy oscillations in pressure (the thrust becomes
strongly negative), while those for the two gas layers fall
off slowly and steadily to the value unity. This brings out
an essential distinction between the two types of cushioning
devices: the mechanical spring is able to exert a backward
pull on the end of the duct, even though the gas pressure
ratio is well above unity, while such an action is impossible
with a gas used as spring. Also, a gas layer would be likely
to be always a rather "stiff" spring -- for the present purpose
this "stiffness" was assumed to be, on the average, about four
times as large as in the cases treated in the preceding section
(in which κ had the value 0.02.)

In Figure (10.7) the pressure in the <u>second</u> layer of
gas is compared with the gas pressure on the piston in the
mechanical spring case. As one sees, the curves in the cases
to be compared follow each other very closely. In the cases
with pressure ratios of 31.6 and 125 even the slight
irregularities near $\tau = 0.1$ occur for both mechanical spring
and gas layer.

FIG. 10.1   CONFIDENTIAL   123

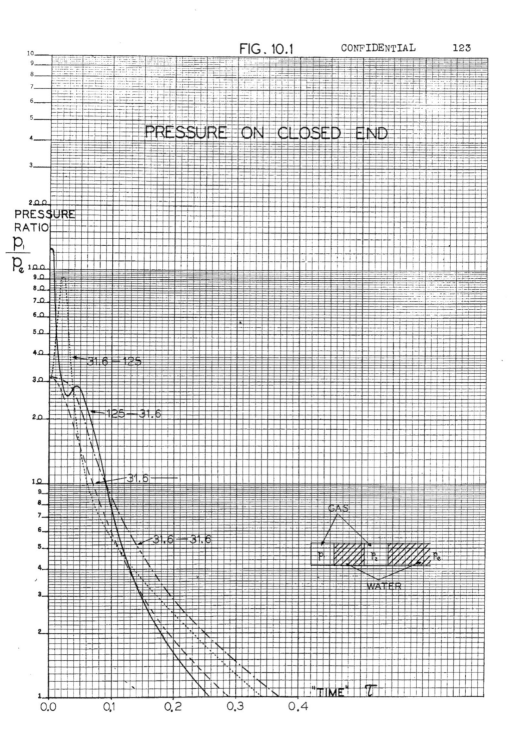

FIG. 10.2   CONFIDENTIAL   125

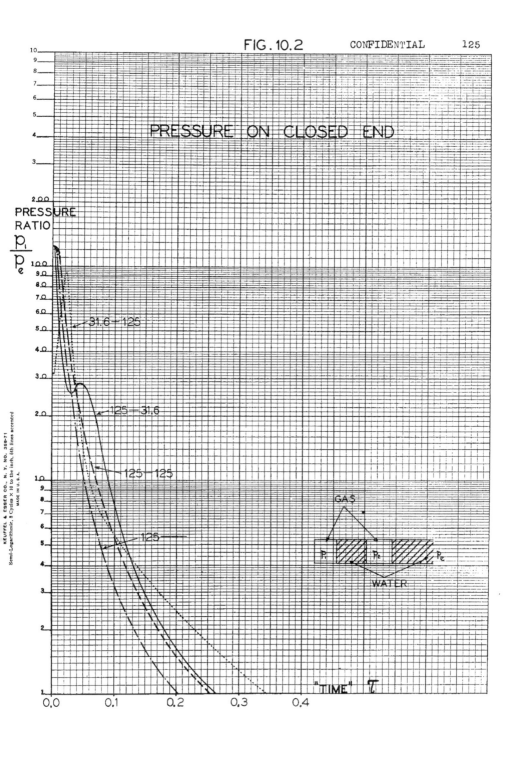

FIG.10.3   CONFIDENTIAL   127

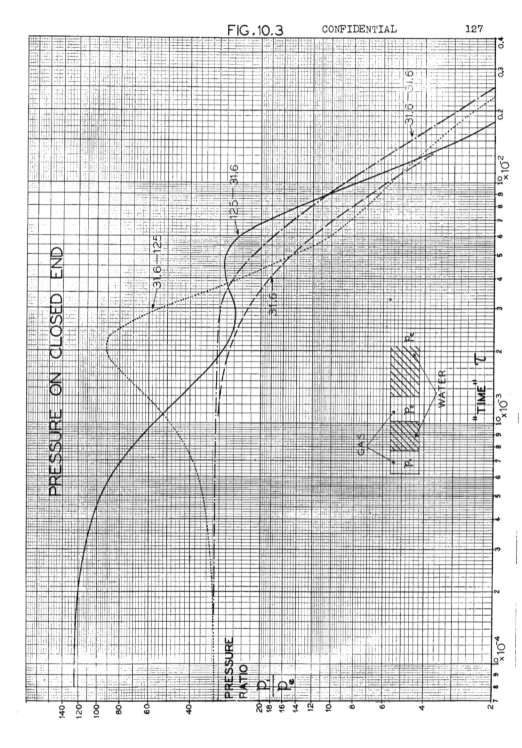

PRESSURE ON CLOSED END

FIG.10.4   CONFIDENTIAL   129

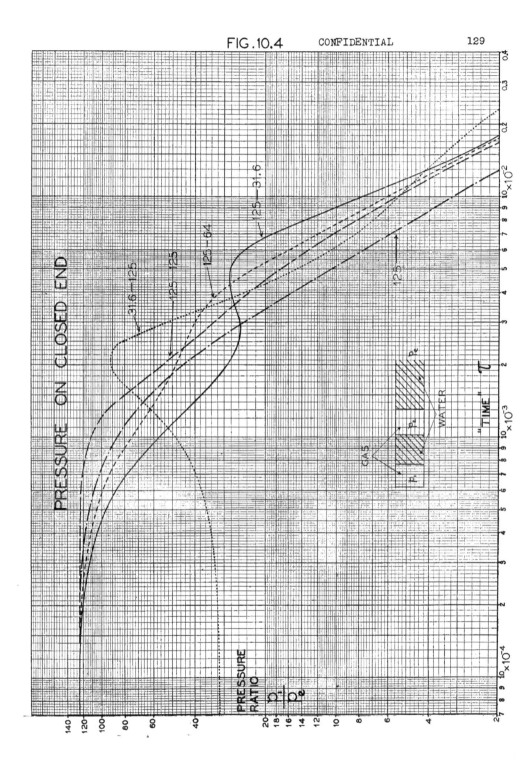

FIG 10.5   CONFIDENTIAL   131

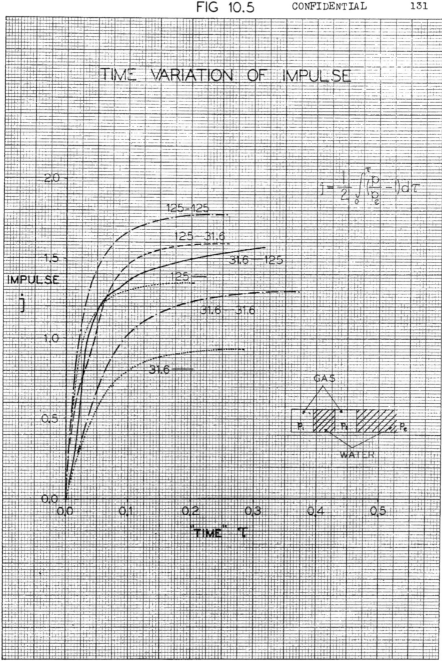

FIG.10.6 . CONFIDENTIAL 133

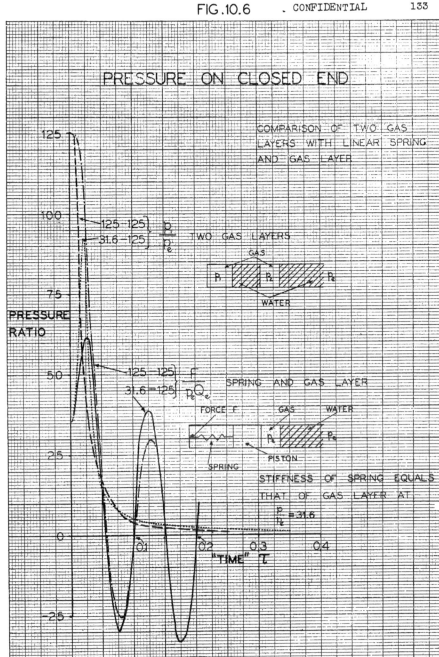

PRESSURE ON CLOSED END

FIG. 10.7   CONFIDENTIAL   135

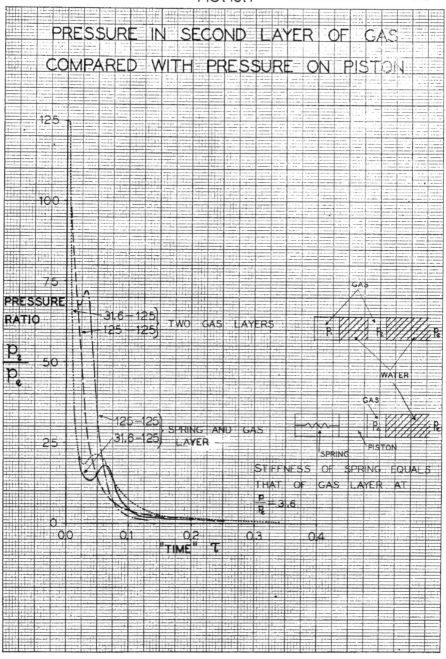

PRESSURE IN SECOND LAYER OF GAS
COMPARED WITH PRESSURE ON PISTON

Section 11. The valveless hydropulse.*

Except in this and the following section we assume
always that the left end of the hydropulse tube is closed by
valves of some kind. In this section we consider the possibility
of operating a hydropulse without valves.

We shall study the simple device indicated in the figure:

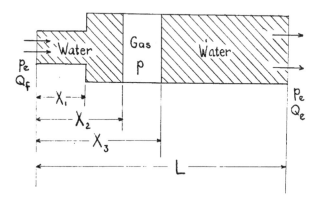

We assume that a layer of gas is injected into the tube while
water is flowing through it, toward the right, as indicated
in the figure. A constriction of the tube is provided in front
of the gas layer, so that a thrust becomes possible. Our
object is to study the motion during the time required for
the gas to expand down to the outside pressure p_e. In other
words, we consider here only what occurs during a single stroke.**

---

* The idea of treating this problem was suggested to us by
   Ensign L. B. Edelman, of the Bureau of Aeronautics Project,
   Annapolis, Md.

** The hydropulse without valves should be distinguished from
   the hydroduct (F. Zwicky's terminology.) in the hydroduct
   the gas would be introduced <u>continuously</u>, not intermittently,
   so that a steady flow would result.

We make the same assumptions regarding the behavior of the water and gas layers as in Section 2. We denote the volume of the wide section of the tube, i.e. that extending from X_1 to the exit section, by V_e and introduce the following notation for the relative volumes of the various water and gas layers:

$$w_1 = \frac{X_1 Q_f}{V_e} \ , \ w_2 = \frac{(X_2 - X_1)Q_e}{V_e} \ , \ w_3 = \frac{(L - X_3)Q_e}{V_e}$$

(11.1)

$$v = \frac{(X_3 - X_2)Q_e}{V_e} \ .$$

We denote by q the ratio of the cross-section areas at the front and exit sections:

(11.2) $$q = Q_f/Q_e \ .$$

The equations of motion are then readily derived, essentially as in Section 3, with the result:

(11.3) $$(1 - w_2 - w_3)^{-\gamma} v_s^{\gamma} - 1 = -2w_3'' w_3 \ ,$$

(11.4) $$(1 - w_2 - w_3)^{-\gamma} v_s^{\gamma} - 1 + (w_2')^2 (1 - 1/q^2)$$

$$+ 2w_2''(w_2 + w_1/q^2) = 0.$$

The dependent variables are w_2 and v: w_1 is a constant and w_3 is determined at once when w_2 and v are known, since $w_2 + v + w_3 = 1$. The quantity v_s is the value of v when the pressure in the gas layer has the value p_e, and γ is the adiabatic exponent; that is, $p_e v_s^{\gamma} = pv^{\gamma}$. The primes refer to differentiations with respect to the dimensionless time τ. The initial conditions are:

$$(11.5) \qquad w_2\big|_{\tau=0} = w_{21} \;,\quad v\big|_{\tau=0} = v_1 \;,$$

$$w_2'\big|_{\tau=0} = w_{21}' \;,\quad v'\big|_{\tau=0} = v_1' \;,$$

that is, we prescribe in effect the initial positions and velocities of the gas-water interfaces.[*]

The number of parameters involved here is rather large: we have been forced to restrict ourselves to a few of the many possibilities. In all cases we have taken $v_s = 0.20$, $w_{21} = 0.20$, $v_1' = 0$, $q = 0.25$, and $\gamma = 1.5$. This means that the expanded volume of the gas (when at pressure p_e) and the initial volume of the water layer extending from the constriction to the gas layer are each 1/5 of the volume of the wider portion of the tube. The assumption that $v_1' = 0$ means that the initial velocities of the two gas-water interfaces are assumed to be the same. In addition the cross-section areas of the two different portions of the tube are assumed to be in the ratio 1:4. With respect to the remaining parameters we have considered the following four cases:

1) $v_1 = .008$ (or, $p_1/p_e = 125$), $w_{21}' = 0$, $w_1 = 0.20$

2) $v_1 = .008$ (or, $p_1/p_e = 125$), $w_{21}' = 0.279$, $w_1 = 0.20$

3) $v_1 = .008$ $(p_1/p_e = 125)$, $w_{21}' = 0.279$, $w_1 = 0.025$

4) $v_1 = .020$ $(p_1/p_e = 31.6)$, $w_{21}' = 0.279$, $w_1 = 0.025$

The initial pressure ratios p_1/p_e result from the adiabatic

---

[*] The dimensionless velocity of the left interface, namely $\frac{d}{d\tau}\left(\frac{X_2}{L-X_1}\right)$, is equal to w_2'; that of the right interface is $-w_3'$.

relation (with $\gamma = 1.5$). In Cases 1) and 2) the volume of
the constricted forward part of the tube is 1/5 that of the
wider part, and 1/40 in Cases 3) and 4). The value $w'_{21} = 0.279$
means that water would be flowing into the entrance section
of the tube at a rate of about 40 ml./hr. if p_e were 1 atm.
In Case 1) the initial velocity is zero. Two different initial
pressure ratios are also taken.

In each of the above cases, the solutions of (11.3) and
(11.4) were obtained by step-wise numerical integration[*]-- a
simple, but laborious, process. The results are given in the
form of graphs in Figures (11.1), (11.2), and (11.3). Figure
(11.1) shows the variation in the (dimensionless) impulse j,
given by

$$(11.6) \qquad j^{(\tau)} = \frac{1}{2} \int_0^T (\frac{p_1}{p_e} - 1) \, d\tau \times (1 - q)$$

The quantity p_1 is the pressure at the shoulder. The curves
for each of our four cases are shown, together with two cases
of the hydropulse with a valve (i.e. with the tube closed off)
located at the position where the left hand gas-water interface
is located initially in the case of the valveless hydropulse.
The curves for Cases 1) and 2) fall too close together to be
distinguishable on our graph, which means that the impulse is
hardly affected at all by a quite considerable change in initial
velocity of the water. The corresponding curve for the hydropulse with a valve (and initial velocity of the water the same
as in Case 2)) shows the total impulse to be greater with a
valve. A considerable decrease in length of the narrower forward
part of the tube (in the ratio of 8 : 1) leads to a sharp loss
of impulse, as we see by comparing the curve for Case 3) with
that for the Cases 1) and 2). This is, of course, what one

---

[*] See Appendix I for the details.

expects: some impulse is lost because of the transport of
momentum out of the front end of the tube. The curve for Case 4),
in which the initial pressure ratio is 31.6 (instead of 125,
as in the other cases) and in which also the narrow front
section is short, indicates a considerable loss in impulse by
comparison with the curve for the corresponding case of a hydropulse closed off at the left end.

Figure (11.2) gives graphs of the pressure ratio in the
gas layer and at the shoulder where the tube broadens out.
The latter quantity p_1/p_e was determined from the relation

$$(11.7) \qquad p_1/p_e = p/p_e + 2\ddot{w}_2\, w_2.$$

The pressure on the shoulder (which of course produces the
thrust) thus will be nearly that of the gas if $\ddot{w}_2$ is small --
that is, if the acceleration of the water in front of the
shoulder is small. The graphs of Figure (11.2) indicate that
the two pressures do not differ greatly, except at the beginning
of the stroke in Case 3), in which the "nose" is short and the
initial gas pressure high.

It is, in fact, possible to draw a few generally valid
(and physically quite reasonable) conclusions from (11.7) and
(11.4), with regard to the effect of varying the front or "nose"
of the tube. We see that the coefficient of $\ddot{w}_2$ in (11.4)
becomes large when w_1 is made large or q is made small, that
is, when the relative volume of the nose is made large or its
relative cross-section area is made small. Both of these
changes, in other words, lead to large values of the "effective
inertia" of the water in the front part of the tube. In these
cases, then, $\ddot{w}_2$ can be expected to be small. Thus a long,
narrow nose will in general lead to a high thrust, since then
p_1/p_e will differ only slightly from p/p_e and, in addition,
the shoulder area on which p_1 acts to create the thrust is
larger.

In Figure (11.3) the velocities of the gas-water
interfaces are shown. The results are what one expects. The
sharper variations in velocity are correlated with the higher
initial gas pressures and shorter length of nose. We observe
also that injection of the gas at a pressure ratio of 125 is
just sufficient to bring the water in the front part of the
tube (which has initially a velocity of the order of 40 mi./hr.)
to rest.

Perhaps the main conclusion to be drawn from the analysis
of this section is the following: As far as the total impulse
for a single stroke is concerned, we see that a hydropulse
without valves could be designed so that its performance would
compare quite favorably with that of the valved hydropulse.
To do this, however, a fairly long narrow forward or nose section
would seem to be needed. As we shall see in the next section,
such a design is likely to cause grave difficulties in the
"scavenging" part of the cycle, during which the tube should
be refilled after the power stroke.

FIG. 11.1   CONFIDENTIAL   143

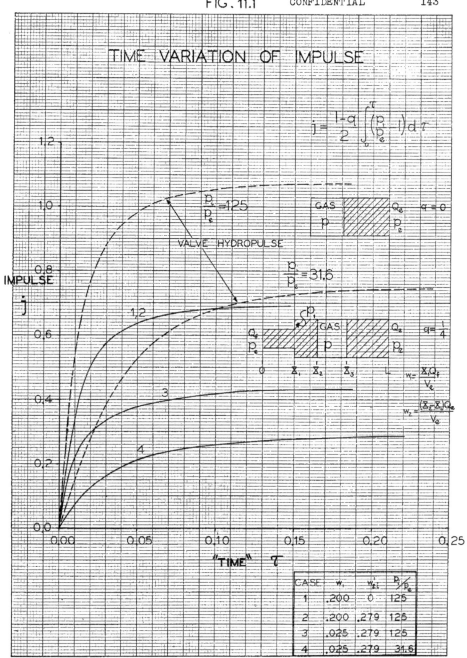

TIME VARIATION OF IMPULSE

$$j = \frac{1-q}{2} \int_0^\tau \left(\frac{p}{p_e} - 1\right) d\tau$$

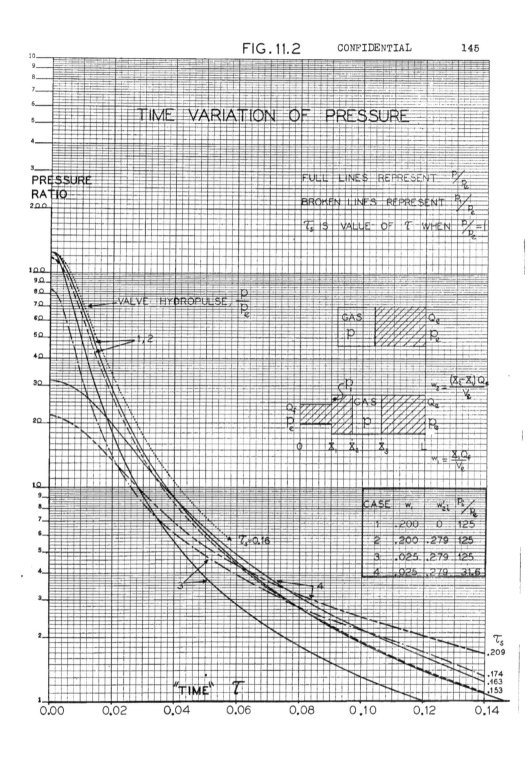

FIG. 11.2   CONFIDENTIAL   145

TIME VARIATION OF PRESSURE

PRESSURE RATIO

FULL LINES REPRESENT $\frac{p}{p_e}$

BROKEN LINES REPRESENT $\frac{p_1}{p_e}$

τ_s IS VALUE OF τ WHEN $\frac{p}{p_e}=1$

VALVE HYDROPULSE, $\frac{p}{p_e}$

1,2

$w_2 = \frac{(\bar{x}_3 - \bar{x}_2)Q_e}{V_e}$

$w_1 = \frac{\bar{x}_1 Q_f}{V_e}$

$\tau_s = 0.16$

3

4

CASE	w_1	w'_{2i}	p_e/p_e
1	.200	0	125
2	.200	.279	125
3	.025	.279	125
4	.025	.279	31.6

τ_s
.209
.174
.463
.153

"TIME" τ

FIG 11 3   CONFIDENTIAL   147

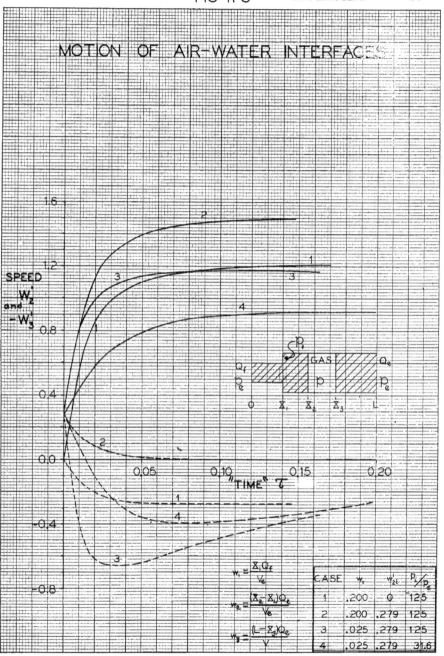

MOTION OF AIR-WATER INTERFACES

Section 12. Scavenging.

In this report we have in general concentrated our
attention on a single exhaust stroke of the hydropulse and have
left aside problems concerning other phases of the full cycle,
in particular, the phase of "scavenging." By this is meant
the portion of the cycle during which the tube is being refilled
with water in preparation for the injection of fuel and the
beginning of a new exhaust stroke. In the present section our
aim is to make a study of the scavenging process in the simplest
possible case in order to get at least a qualitative idea of
what to expect. As we shall see, the results indicate that
the scavenging phase of the cycle may well present problems of
crucial importance for the hydropulse development as a whole.

We consider the simplest possible model of a hydropulse
with a valve: a straight tube with a bank of valves inside
the tube, as indicated in the figure.

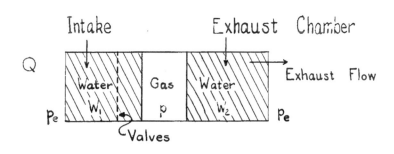

The problem we wish to consider is the following: We assume
that the exhaust stroke has just been completed, i.e. that the
gas has expanded (with the valves closed) until its pressure
has decreased to the value p_e on the outside of the tube and
the water in the exhaust chamber has acquired a certain velocity.
An instant later the gas will be at a pressure less than p_e,
(since the inertia of the moving mass of water will lead to a
further expansion of the gas), so that valves could be opened
automatically by the pressure of the water in the forward part
of the tube. Our purpose in this section is to study the
subsequent motion of both layers of water, on the basis of our
usual assumptions regarding the <u>behavior</u> of gas and water layers.
In particular, we assume the exterior pressure p_e to remain
at the same constant value at both ends of the tube throughout
the motion. In addition, we assume that the valves have a
negligible effect on the flow once they have been opened.

We denote by V_e the total volume of the tube in the
"exhaust chamber" to the right of the valves. The quantities
W_1 and W_2 represent the volumes of water at any instant in the
"intake" and "exhaust" sections of the tube respectively (see
the figure), while W_3 represents the volume of gas at any
instant. These quantities are therefore all functions of the
time. It is also convenient to introduce the quantity W_0 for
the volume of the tube to the left (i.e. in front of) the
valves, which means that W_0 is the value of W_1 at the initial
instant. The equations of motion are obtained, in the usual
way, in terms of the relative volumes $w_i = W_i/V_e$, with i
ranging from 0 to three. They are:

$$(12.1) \quad \begin{cases} 2w_1 w_1'' = 1 - (1 + w_0 - w_1 - w_2)^{-\gamma} \, w_{3s}^{\gamma} \\[2ex] 2w_2 w_2'' = 1 - (1 + w_0 - w_1 - w_2)^{-\gamma} \, w_{3s}^{\gamma} \end{cases}$$

in which the primes mean, as always, differentiations with respect to the dimensionless time τ and the relative volume w_3 of the gas layer has been replaced by

$$(12.2) \qquad w_3 = 1 + w_0 - w_1 - w_2 \; .$$

w_{3s} is the value of w_3 when $p = p_e$. Since w_0 is a constant, the two equations (12.1) serve to determine the two quantities w_1 and w_2, once appropriate initial conditions have been prescribed.

We report here the results of integrating (12.1) in the following special case: We assume the volume of the tube in front of the valves to be the same as w_{3s}, that of the gas layer when the latter is at the exterior pressure p_e, while the exhaust chamber volume V_e is four times as great. (In other words, the exhaust chamber is four times as long as the intake section.) Initially, the water in front of the valves is at rest, while that in the exhaust chamber has the velocity acquired during the previous expansion of the gas layer. The latter initial velocity was obtained from previous calculations (discussed at the end of Section 7) on the assumption of an initial pressure ratio p_i/p_e in the gas layer of 125 and 1.5 as the value for the adiabatic exponent γ. As initial conditions for (12.1) we have therefore:

$$(12.3) \qquad \begin{cases} w_1(0) = w_{1i} = \frac{1}{4} \\ \\ w_1'(0) = 0 \qquad \text{and} \end{cases}$$

$$(12.4) \qquad \begin{cases} w_2(0) = 1 - w_{3i} \\ \\ w_2'(0) = w_{2i}' \; , \end{cases}$$

CONFIDENTIAL

in which w_{31} and w'_{21} are values obtained from earlier calculations for the case of an initial gas pressure ratio of 125.[*]

 The accompanying graph shows the variation of w_1 as a
function of the time τ, that is, it indicates how the tube
refills after the valves open, i.e. how the scavenging phase
takes place. The numerical integration was carried out up to
the point where the pressure in the gas layer had reached its
minimum. The value of w_2 as a function of τ was also obtained,
but is not given here.

 The essential conclusion to be drawn from the graph is
that the refill process is very slow. As indicated on the
graph, the time of stroke τ_s required for the gas to expand
from a pressure p = 125 p_e down to P = p_e with the valves
closed is τ_s = 0.205. When the valves are opened and the
refill flow (or scavenging) takes place we observe that only
1/5 of the exhaust chamber has been refilled during a time
of about τ = 0.6. In other words, the scavenging operation
(as we have considered it here) would require three times as
much time as the power stroke and <u>even then</u> the refilled
volume would be less than that of the gas layer at the instant
of opening the valves.

 That the above conclusion was to be expected on physical
grounds is clear: The mass of water in front of the valves
is at rest when the valves are open. Once the valves are
opened, this mass begins to move under the pressure difference
existing between the gas pressure and that in the water in
front of the tube. Initially this difference is very small,

---

[*] A few calculations were made in other cases also, but we
do not report on them here in detail. However, one result
may be of interest. It is that the flow into the exhaust
chamber is not greatly affected if it is assumed that the
valves open when the gas pressure has dropped to a value of
0.9 p_e instead of p_e.

CONFIDENTIAL

so that the velocity of the mass of water increases very
slowly at first (see the attached graph). The pressure
difference is built up only as the gas layer over-expands
due to the motion of the water in the exhaust chamber, and
this appears to be a rather slow process.

It seems certain that efficiency of operation of the
hydropulse requires that the scavenging phase should not take
up too large a fraction of the entire cycle of operations.
For the very simple design of a hydropulse discussed above,
it is then clear that the section of the tube in front of the
valves should be eliminated entirely so that the refill would
take place as quickly as possible: Clearly, the smaller the
initial mass in front of the valves the greater the acceleration
at the beginning of the process and the more rapid the entry
of water. However, if the front section of the tube is very
short, or nonexistent, our assumption that the pressure over
the open end is constant is likely to be quite wrong. To
obtain reasonable accuracy in such cases it would be necessary
to consider the flow pattern near the end of the tube in a
detailed way -- both theoretically and experimentally. The
conclusion that the part of the tube in front of the valves
should be eliminated entirely to aid in efficient scavenging
is just the reverse of the conclusion reached in the preceding
section, where the operation of the valveless hydropulse was
considered. There we found that a long constricted front
portion of the tube was favorable for the operation of the
valveless hydropulse, principally because a high effective
inertia for the mass of water in the front part of the tube
would be obtained. For scavenging, just the reverse effect
is desired, which leads us to suspect that the front section
of the tube ought to be flared out to promote scavenging, as
indicated on the figure:

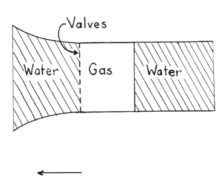

Direction of motion

Another of our basic assumptions which may be seriously at fault under the conditions of the scavenging process is that referring to the existence of a <u>layer</u> of gas with <u>plane</u> interfaces across the whole section of the tube. Once the gas pressure has dropped down to about the exit pressure, the actual configuration of the gas volume may differ radically from that of a slice. It seems possible that the gas volume may break up into bubbles clinging to the top of the tube, and that gas might then escape through the front end of the tube, especially if the valves are close to the front. This is a question which should be dealt with experimentally.

Of course, the very rudimentary design upon which our above remarks are based is not the only possible one. In fact, our discussion indicates that the kind of "passive" scavenging encountered in our simple model should be replaced by some sort of "forced" scavenging. That is, part of the energy in the gas should be used to increase the rate of flow during scavenging. The turbo-jet offers an example of a

solution to this problem for duct devices in air, though
here the purpose of the turbine is more to create a high
pressure in the combustion chamber for the sake of thermo-
dynamic efficiency than to increase the flow in order to
increase the thrust-augmenting mass. Just what means should
be taken to accomplish similar results for the hydro-devices
is far from clear at the present stage of development of these
devices. Perhaps valves distributed along the sides of the
hydropulse tube might be used to facilitate the refill process.
Another method which might provide forced scavenging is the
following: Two tubes placed side by side would "fire" into
a common nozzle, the tubes being fired alternately. The
high velocity and consequent low pressure produced in the
nozzle by the firing of one tube would then facilitate the
scavenging of the other.

FIG. 12.1   CONFIDENTIAL   157

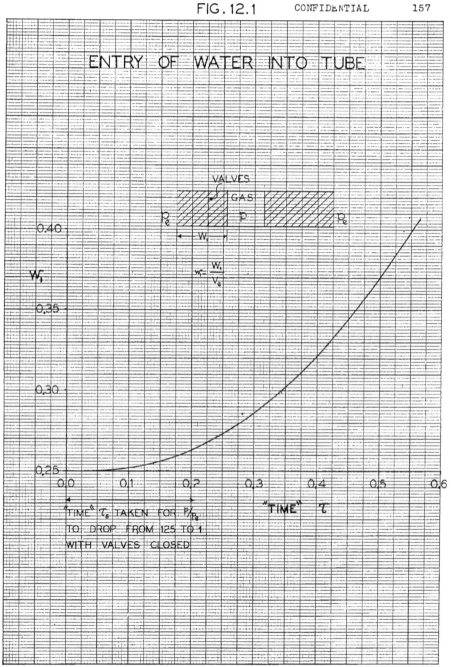

ENTRY OF WATER INTO TUBE

Section 13. Effect of finely distributed gas bubbles.

In actual practice it may be that the high pressure gases will not be introduced as a single layer but rather in the form of a large number of small bubbles thoroughly mixed in the water. For example, this might occur if the chemical which reacts with the water to form the gas were to be thrown into the water as a finely ground powder. It is of interest to know how this might influence the behavior of the hydro-pulse, as compared with the behavior when the gas is in a single layer.

The following assumptions will be made:

1) The gas bubbles are distributed uniformly in the water.

2) The gas bubbles move together with the surrounding water.

3) The initial velocity of the water in the tube is zero.

In other words, the mixture of gas bubbles and water will be treated as homogeneous.

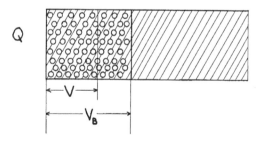

We shall show that such a mixture of gas bubbles and
water is equivalent as far as the behavior of the hydropulse
is concerned to a single gas layer with a different total
volume of the tube. More precisely, let V_W be the volume of
that part of the water which contains the gas bubbles. Then
only 1/3 of V_W is effective as far as thrust augmentation is
concerned. That is, if V_e is the total volume of water in the
tube, the equivalent tube with a single gas layer should have
the volume

$$V_e - \frac{2}{3} V_W \, ,$$

instead of $V_e - V_W$. The reason for this is that the mixture
of gas bubbles and water does not partake of the full velocity
of the clear water, but has a velocity which varies linearly
from zero at the front end to the velocity of the interface.
Of course, the factor 1/3 is to be considered as a rough
estimate, since any deviations from the assumptions 1), 2)
and 3) would change it.

At any time t, let V_G be the total volume of the gas
included in all the bubbles, and let V_B be the volume of the
gas-water mixture. Since the total amount of water in the
mixture remains constant, by assumption 2), we have (see
the accompanying figure):

(13.1) $$V_B = V_G + V_W.$$

Furthermore, the density ρ_B of the mixture is very closely
given by

(13.2) $$\rho_B = \rho \, \frac{V_W}{V_B} \, ,$$

in which ρ is the density of the water.

Let us compute the velocity of the mixture at any cross-section, enclosing with the front end the volume V. At the front end, the velocity of the mixture is 0; at the interface between the mixture and the clear water, the velocity is

$$\frac{\dot{V}_G}{Q} ,$$

in which $\dot{V}_G$ is the time rate of expansion of the total volume of gas (in all bubbles) and Q is the cross-section area of the tube, assumed constant. Assumption 3) is used at this point. We proceed to show that the velocity of the mixture varies linearly with V: The rate of increase of the volume of the gas enclosed between this cross-section and the front end is

$$\dot{V}_G \cdot \frac{V}{V_B}$$

and thus is equal to QU where U is the velocity of the mixture of the section characterized by the volume V. Thus

(13.3) $$U = \frac{\dot{V}_G}{Q} \cdot \frac{V}{V_B}$$

which varies linearly with V.

The kinetic energy of the mixture is

$$\int_0^{V_B} \frac{1}{2}\rho_B U^2 dV = \frac{1}{2}\rho \frac{V_W}{V_B} \cdot \left(\frac{\dot{V}_G}{Q V_B}\right)^2 \int_0^{V_B} V^2 dV = \frac{1}{2}\rho \frac{\dot{V}_G^2}{Q^2} \cdot (\frac{1}{3} V_W).$$

The kinetic energy of the clear water is

$$\frac{1}{2}\rho \frac{\dot{V}_G^2}{Q^2} (V_e - V_B)$$

The total kinetic energy, making use of (13.1) is therefore

(13.4)
$$\frac{1}{2}\rho \frac{\dot{V}_G^2}{Q^2} \left(V_e - \frac{2}{3} V_W - V_G \right).$$

Expression (13.4) is identical with the kinetic energy in the case of a hydropulse with a single layer of gas when the initial velocity of the water in the tube is zero, except that $V_e - \frac{2}{3} V_W$ replaces V_e. Furthermore, the work done by the gas is the same in the two cases since it depends only on the adiabatic law and the change in volume of the gas. The energy equation thus shows that the equivalent hydropulse with a single block of gas has a volume

$$V_e - \frac{2}{3} V_W.$$

Appendix I

The method of finite differences applied to

the numerical solution of hydropulse differential equations.

In certain parts of this report, (notably in Sections 9, 10, 11, and 12), it was necessary to determine numerically the solutions of systems of non-linear differential equations under prescribed initial conditions. Further extensions and refinements of the hydropulse theory may well require the integration of still more complicated systems than those we have considered hitherto. Consequently, it was thought worth while to include a brief discussion of the methods we used and found convenient for this type of problem, even though the general method of numerical integration by using finite differences is very well known.

The differential equations of motion which appear in Sections 9, 10, and 12 have the general form:

$$\frac{d^2u}{dx^2} = f(u,v)$$

(I.1)

$$\frac{d^2v}{dx^2} = g(u,v) \ ,$$

with initial conditions (i.e. conditions at $x = x_1$):

(I.2)
$$u\big|_{x=x_1} = u_1 \qquad v\big|_{x=x_1} = v_1$$

$$\frac{du}{dx}\Big|_{x=x_1} = u_1' \qquad \frac{dv}{dx}\Big|_{x=x_1} = v_1'$$

in which u_1, u_1', v_1, and v_1' are given constants. We wish to

determine u and v as functions of x. It is possible to solve
equations (I.1) numerically by replacing them with appropriate
difference equations. This can be carried out in the following
way:

Let us consider one of the dependent variables, u.
The value of u at $x_0 + h$ and $x_0 - h$ can be expressed in terms
of the value of u and its derivatives at $x = x_0$ by a Taylor
series.

$$(I.3a) \qquad u(x_0 + h) = u(x_0) + hu'(x_0) + \tfrac{1}{2}h^2 u''(x_0)$$
$$+ \tfrac{1}{3!} h^3 u'''(x_0) + \tfrac{1}{4!} h^4 u^{IV}(x_0) + \ldots$$

$$(I.3b) \qquad u(x_0 - h) = u(x_0) - hu'(x_0) + \tfrac{1}{2}h^2 u''(x_0)$$
$$- \tfrac{1}{3!} h^3 u'''(x_0) + \tfrac{1}{4!} h^4 u^{IV}(x_0) - \ldots$$

Adding (I.3a) and (I.3b) we have

$$u(x_0 + h) + u(x_0 - h) = 2u(x_0) + h^2 u''(x_0) + \tfrac{1}{12}h^4 u^{IV}(x_0) + \ldots$$

or

$$(I.4) \quad u(x_0 + h) = 2u(x_0) - u(x_0 - h) + h^2 u''(x_0) + \tfrac{1}{12}h^4 u^{IV}(x_0) + \ldots$$

If h is taken small enough, it is possible to neglect all
terms of order higher than the second in h, yielding

$$(I.5) \qquad u(x_0 + h) = 2u(x_0) - u(x_0 - h) + h^2 u''(x_0)$$

If the values of u and v are both known for $x = x_0$, then we can use equation (I.1) to obtain $u''(x_0)$:

$$u''(x_0) = f(u[x_0], v[x_0]) = f(x_0)$$

and hence (I.5) can be written in the form

(I.6) $\qquad u(x_0 + h) = 2u(x_0) - u(x_0 - h) + h^2 f(x_0)$,

with $f(x_0)$ known.

If we take $x_0 = x_1$ in (I.3a) we can determine $u(x_1 + h)$ and, similarly, $v(x_1 + h)$ with the help of equations (I.1) and (I.2) thus

$$u(x_1 + h) = u_1 + h u_1' + \frac{1}{2} h^2 f(u_1, v_1) ,$$

(I.7) and

$$v(x_1 + h) = v_1 + h v_1' + \frac{1}{2} h^2 g(u_1, v_1) ,$$

in which derivatives of order higher than the second are ignored.[*] It is now possible to determine $u(x_1 + 2h)$ and $v(x_1 + 2h)$ from equation (I.6) and the corresponding equation for v, by taking $x_0 = x_1 + h$. Taking x_0 in succession equal to $x_1 + 2h$, $x_1 + 3h, \ldots$, $x_1 + nh, \ldots$ it is possible in this way to calculate approximate values for u and v as functions of x at points which are a distance h apart.

Any integral of a function of u or v, say

$$I = \int_a^b \phi(u) dx$$

can be determined quite simply with the help of Simpson's

---

[*] One could include higher order terms. For this it would only be necessary to calculate higher derivatives of u at $x = x_0$ by differentiating (I.1).

CONFIDENTIAL

rule, using intervals which are some integral multiple of
h in length, since the values of u, and hence of ϕ (u), are
known at the ends of all such intervals.

In order to expedite the integration of (I.1) it is
desirable to use the largest value of h consistent with the
accuracy which is demanded in any particular instance. The
largest term in equation (I.4) which is neglected is $\frac{1}{12}h^4 u^{IV}(x_0)$
and this will be the approximate error in any one step.
This error can be estimated in a simple manner in the course
of the numerical calculations from the following considerations.
Equations (I.3), (I.4) and (I.5) can be written for u'' rather
than u as the function of x. We get

(I.5') $\quad u''(x_0 + h) = 2u''(x_0) - u''(x_0 - h) + h^2 u^{IV}(x_0)$

or

(I.8) $\quad \frac{1}{12}h^4 u^{IV}(x_0) = \frac{1}{12}\ h^2 u''(x_0 + h) - 2h^2 u''(x_0) + h^2 u''(x_0 - h)$

$$= \frac{1}{12}\ h^2 f(x_0 + h) - 2h^2 f(x_0) + h^2 f(x_0 - h) \quad ,$$

which is $\frac{1}{12}$ of the fourth difference of u. Thus, by taking a
running second difference of $h^2 f(x)$ we obtain an estimate of
the error in any step. If this error is much smaller than
necessary we may increase h and carry out the integration more
rapidly. If the error is too large, h should be decreased.

Any such change in h from h_1 to h_2 during the calculations, say at x = a, involves an interpolation procedure
to determine u at $x = a - h_2$ before (I.6) can be used again.
This interpolation can be carried out in several ways, for
example, by the use of (I.3b) with suitable estimates for
$u'(a)$, etc. (It should be noted that the integration of u
and v must be carried out simultaneously, and the same h should

be used for both integrations.)

The following numerical example will illustrate the way in which the choice of h affects results:

We consider the case of two gas layers, as described in Section 10. The equations of motion for the system were given as

(I.9)
$$
\begin{cases}
\xi_s^{\gamma}\xi^{-\gamma} - \eta_s^{\gamma}\eta^{-\gamma} = 2\alpha\xi'' \\[2mm]
\eta_s^{\gamma}\eta^{-\gamma} - 1 = 2(\xi'' + \eta'')(1 - \xi - \eta - \alpha)
\end{cases}
$$

or

(I.10)
$$
\begin{cases}
\xi'' = \dfrac{1}{2\alpha}(\xi_s^{\gamma}\xi^{-\gamma} - \eta_s^{\gamma}\eta^{-\gamma}) \\[3mm]
\eta'' = \dfrac{1}{2(1-\xi-\eta-\alpha)}[\eta_s^{\gamma}\eta^{-\gamma} - 1] - \dfrac{1}{2\alpha}(\xi_s^{\gamma}\xi^{-\gamma} - \eta_s^{\gamma}\eta^{-\gamma})
\end{cases}
$$

We consider the case in which the parameters have the following values:

(I.11) $\qquad \gamma = 1.5$, $\quad \xi_s = \eta_s = \alpha = 0.25$,

and as initial conditions, we take

(I.12) $\quad \xi_1 = 0.01$, $\quad \eta_1 = 0.025$, $\quad \xi_1' = \eta_1' = 0$ for $\quad \tau = 0$.

Substituting in (I.9) we obtain

(I.13) $\qquad \xi_1'' = 186.76$, $\quad \eta_1'' = -165.35$

and (I.7) becomes

$$\xi(\Delta\tau) = .01 + 93.38(\Delta\tau)^2 - 3.1795 \times 10^5(\Delta\tau)^4 + \ldots$$

(I.14)

$$\eta(\Delta\tau) = .025 - 82.68(\Delta\tau)^2 + 3.2712 \times 10^5(\Delta\tau)^4 + \ldots$$

In order to make these series converge fairly quickly, we must take $\Delta\tau$ of the order of 0.005 or less. Once $\Delta\tau$ has been chosen, we can proceed as already indicated, with the repeated use of equation (I.6) in which u and v are to be replaced by ξ and η .

The accompanying table shows the results which are obtained for ξ as a function of τ when h (or $\Delta\tau$) takes on the values 0.00125, 0.00250, 0.0050, 0.0100.

It can be seen that h = 0.00125 is smaller than necessary for our purposes, and the accuracy is not appreciably affected by taking, rather, h = 0.0025. It will be found, furthermore, that $\frac{1}{12}$ of the fourth difference is completely negligible in the former case ($\Delta\tau$ = 0.00125) and even over most of the range for h = 0.0025.

In order to illustrate the effect of changing the interval, we have carried this out for the case in which h = 0.0025 initially. The interval is successively doubled at $\tau = 0.0350$ and at $\tau = 0.050$, as the fourth difference is quite small at each of these points. It can be seen that the error introduced as a result is quite small -- of the order of 1 per cent. It should be noted, in connection with the use of equation (I.6) at τ_0 = 0.0350, that when the $\Delta\tau$ interval is doubled we must take

$$\xi(\tau_0 - h) = 0.02870$$

τ	ξ				
	$\Delta\tau=0.00125$	$\Delta\tau=0.0025$	$\Delta\tau=0.005$		$\Delta\tau=0.01$
0.	.01000				
0.00125	1015				
.00250	1058	.01057			
.00375	1127				
.00500	1218	1217		1214	
.00625	1327				
.00750	1450	1449			
.00875	1583				
.01000	1722	1720		1718	.01616
.0125		2001			
.0150		2266		2261	
.0175		2494			
.0200		2670		2657	2565
.0225		2788			
.0250		2853		2827	2718
.0275		2876			
.0300		2870		2837	
.0325		2846			
.0350		2814		2782	
.0375		2781			
.0400		2753	02756	2729	2674
.0425		2735			
.0450		2731	2738	2719	
.0475		2743			
.0500		2773	2784	2775	2809
.0525		2822			
.0550		2890		2907	
.0575		2977			
.0600		3083	3113 ($\Delta\tau=0.01$)	3115	3222

corresponding to h = 0.005, instead of

$$\xi(\tau_o - h) = 0.02846$$

which would have been correct if h had remained unchanged.

From inspection of the table, it seems reasonable to conclude that it would be most efficient to choose h = 0.005 from the outset if an accuracy of about 1 per cent is desired. Similar considerations would determine the most efficient choice of h under different initial conditions.

Appendix II

Special method for calculation of $v(\tau)$.

The method of finite differences outlined in Appendix I above can of course be used to treat simpler cases than those of Sections 9, 10, and 12. In fact, we have applied it to find solutions in the simple case of the straight tube with one layer of gas. However, for these cases the difference method is more laborious and time-consuming than necessary. In this Appendix we outline a method of calculation for the case of the straight tube with one layer of gas which makes it possible to determine accurately the gas volume as a function of the time over the whole stroke in a fraction of the time required by the method of finite differences. That the method is very accurate (with an error less than 1 per cent in general) we know by comparison with the solutions by finite differences.

The differential equation to be integrated is equation (4.9) of Section 4 for the case $q = 1$, as follows:

$$(\text{II.1}) \qquad \frac{d}{dv}(v')^2 = \frac{v_s^\gamma v^{-\gamma}-1}{1-v} \quad ,$$

in which $v' = \frac{dv}{d\tau}$, the variables being in the dimensionless form discussed in Section 4. Since $\frac{d}{dv}(v')^2 = 2v''$ we may write (II.1) in the form

$$(\text{II.2}) \qquad v'' = \frac{v_s^\gamma v^{-\gamma}-1}{2(1-v)} \quad .$$

The differential equation (II.2) is to be integrated under the initial conditions:

$$(\text{II.3}) \qquad \begin{aligned} v(0) &= v_1 \quad , \\ v'(0) &= v_1' \end{aligned}$$

172

By integration of (II.1) we obtain v' as a function of v:

$$\text{(II.4)} \qquad v'^2 = v_1'^2 + \int_{v_1}^{v} \frac{v_s^\gamma \xi^{-\gamma} - 1}{1 - \xi} \, d\xi \, ,$$

in which v_1' is the initial value of v', as above.

We note that

$$\text{(II.5)} \qquad v' = \frac{1}{\frac{d\tau}{dv}} \, , \quad v'' = - \frac{\frac{d^2\tau}{dv^2}}{(\frac{d\tau}{dv})^3} \, .$$

It follows that all derivatives of $\tau(v)$ with respect to v can be obtained by making use of (II.2) since v' can be considered a known function of v by (II.4). In what follows it is convenient for various reasons to determine τ as a function of v rather than v as a function of τ. By τ', τ'', etc. we mean differentiations with respect to v.

The method of approximating the function $\tau(v)$ which we propose is based on our knowledge of the general character of the function: We know that $\tau(v)$ is a nearly linear function of v over an interval $v_1 + \epsilon \leq v \leq v_s$, $\epsilon > 0$ which extends up to v_s (the gas volume corresponding to the exit pressure p_e) but not all the way down to the initial value v_1 of v. Near v_1, in fact, the curve $\tau(v)$ has high curvature. Our method of approximation thus consists of using a series development for τ in powers of $v - v_s$ for all values of v in an interval $v_1 < v_3 \leq v \leq v_s$, in which v_3 is to be suitably determined, together with a different type of approximation valid over the remaining interval from v_1 to v_3. The main difficulty arises in piecing these two different types of representation of the solution together at $v = v_3$. The circumstances are indicated in the accompanying figure, which should be compared with the accurate solutions given by Figure (7.10) (cf. the later parts of Section 7):

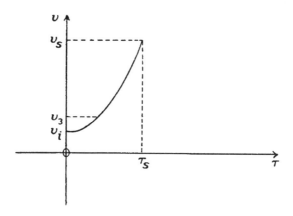

To carry out this program we begin by expanding $\tau(v)$ about $v = v_s$:

$$(II.6) \qquad \tau = \tau_s + \tau_s'(v-v_s) + \frac{\tau_s''}{2!}(v-v_3)^2 + \frac{\tau_s'''}{3!}(v-v_s)^3 + \ldots,$$

with

$$(II.7) \qquad \begin{cases} \tau_s' = 1/v_s' \\[4pt] \tau_s'' = 0 \\[4pt] \tau_s''' = \dfrac{\gamma}{2v_s(1-v_s)} \cdot \dfrac{1}{(v_s')^3} \end{cases}$$

by use of (II.5) and (II.2). The quantity τ_s is, of course, the "time" of stroke. The quantity v_s is known, but τ_s and τ_s' are not known. However, τ_s' can be computed using (II.4), since $v_s' = 1\frac{d\tau}{dv} = 1/\tau_s'$; we need only expand $(1-\xi)^{-1}$ in powers of ξ and then carry out the integration in (II.4) to obtain v'^2 with any desired accuracy as a power series in v. Once this is done the coefficients in the series (II.6) for τ would all be known with the exception of the first one τ_s. To fix the value of τ_s we need to know only one pair of values (v, τ), with v a value not too near v_1 (or, τ not too near zero) so that (II.6) still converges reasonably well.

We next investigate the behavior of $\tau(v)$ near v_1. From $\tau' = \frac{1}{\sqrt{(v')^2}}$ and (II.4) we see that the behavior of $\tau(v)$ near $v = v_1$ depends strongly on the initial value v_1' of v': If $v_1' = 0$, then τ' behaves like $\frac{1}{\sqrt{v-v_1}}$ when v is near v_1, but if $v_1' \neq 0$ then τ' is regular for $v = v_1$. Thus τ would possess two different developments near $v = v_1$, as follows:

(II.8) For $v_1' = 0$: $\tau = a(v-v_1)^{\frac{1}{2}} + b(v-v_1) + c(v-v_1)^{3/2} + \dots$

(II.9) For $v_1' \neq 0$: $\tau = \frac{v-v_1}{v_1'} + \beta(v-v_1)^2 + \delta(v-v_1)^3 + \epsilon(v-v_1)^4 + \dots$

If we were to compute the coefficients a,b,c,β,δ,ϵ by evaluating the successive derivatives of τ at $v = v_1$, we would find the series to converge very badly. Consequently, we break off the developments with the terms in $(v-v_1)^{3/2}$ and $(v-v_1)^4$ and compute the coefficients in the following way: The coefficients a and β are determined so that τ' has the correct value at $v = v_1$, but b and c, δ and ϵ are chosen in such a way

175

that the first derivative τ' has correct values at each of the two points $v = v_2$ and $v = v_3$ given by[*]

(II.10) $\qquad v_2 = v_1 + \dfrac{v_s - v_1}{10}$, $v_3 = v_1 + \dfrac{v_s - v_1}{5}$.

In this way the point v_3 referred to above is located. The values of $\tau'(v_2)$ and $\tau'(v_3)$ are calculated from (II.4) and the relation $\tau = 1/v'$.

$\qquad$ The coefficients a,b,c can now be calculated as follows: One finds readily that

$$\lim_{v \to v_1} v'' = \frac{2}{a^2} ,$$

so that $a = \sqrt{2/v_1''}$. Hence a is fixed since v_1'' is known. Inserting this value for a we then obtain the following pair of linear equations for the determination of b and c:

$$\tau'(v_2) = \frac{1}{2}\sqrt{\frac{2}{v_1''}} (v_2 - v_1)^{-\frac{1}{2}} + b + \frac{3}{2} c(v_2 - v_1)^{\frac{1}{2}}$$

$$\tau'(v_3) = \frac{1}{2}\sqrt{\frac{2}{v_1''}} (v_3 - v_1)^{-\frac{1}{2}} + b + \frac{3}{2} c(v_3 - v_1)^{\frac{1}{2}} .$$

$\qquad$ The calculation of β , σ , and ϵ follows the same scheme.

$\qquad$ The formulas (II.8) and (II.9) are found to be quite accurate approximations to $\tau(v)$ in the interval $v_1 \leq v \leq v_3$. By substituting (v_3, τ_3) in (II.6) an accurate value for τ_s

---

[*] It is clear that v_2 and v_3 need not be chosen exactly in this way. We found that these choices for v_2 and v_3 gave good results.

can be found -- since (II.6) converges quite well for $v \neq v_3$.
Once the value of τ_s is known, (II.6) in the interval
$v_3 \leq v \leq v_s$ together with (II.8) or (II.9) in the interval
$v_1 \leq v \leq v_3$ constitute very good approximations to $\tau(v)$ over
the whole range $v_1 \leq v \leq v_s$.

The method has been checked in a considerable number of
cases by comparison with the results obtained by finite
differences, with results that indicate the error in $\tau(v)$ to
be less than 1 per cent of τ_s. The method of approximation
outlined here can be carried through by an experienced
calculator in a specific case in less than one-half hour, while
the difference method requires about four hours.

Date Due

APR 15 '63			

PRINTED IN U. S. A.

NYU
AMG-
128

Stoker

**N. Y. U. Institute of
Mathematical Sciences**
25 Waverly Place
New York 3, N. Y.